FRAGMENTATION IN AGRICULTURAL LANDSCAPES

Proceedings of the third annual conference of IALE(UK), the UK region of the International Association for Landscape Ecology, held at Myerscough College, Bilsborrow, Preston on 13-14th September 1994.

Edited by: John Dover

IALE International Secretariat
Dr Almo Farina
Lunigiana Museum of Natural History,
54011 Aulla,
Italy

IALE(UK)
Dr Peter Dennis
c/o MLURI
Craigiebuckler,
Aberdeen AB9 2QJ,
UK.

British Library Cataloguing-in-Publication Data
A catalogue record for this book is available
from the British Library

ISBN 0 9524263 0 7

Published by IALE(UK)
Printed in Great Britain by Colin Cross Printers Ltd, Garstang

Contents

Page

PART III: POSTER PAPERS

Delegate List

G. Arnold	CSIRO, Division of Wildlife & Ecology, LMB 4, Midland, Perth, Western Australia 6056.
D. Baldock	Institute for European Environmental Policy, 158 Buckingham Palace Rd., London SW1W 9TK.
S. Bell	The Forestry Commission, 231 Corstorphine Rd., Edinburgh EH12 7AT.
P. Bellamy	Institute of Terrestrial Ecology, Monks Wood, Abbots Ripton, Huntingdon PE17 2LS.
E. Bignal	Kindorchaid, Bruichladdich, Isle of Islay, Argyll PA44 7PT.
R. Bunce	Institute of Terrestrial Ecology, Merlewood Research Station, Windermere Rd., Grange-over-Sands, LA11 6JU
J. Boothby	School of Social Science, Liverpool John Moores University, Trueman Building, Webster St., Liverpool L3 2ET.
R. Buxton	Manor House, Little Wittenham, Abingdon, Oxon OX14 4RA.
S.A. Clarke	Derwent, Cowesfield Gate, Whiteparish, Salisbury SP5 2RB.
A. Cooper	Department of Environmental Studies, University of Ulster, Coleraine, Co. Londonderry, BT52 1SA.
N. Corker	Broomfield College, Morley, Derby DE7 6DN.
D. Crook	University of Huddersfield, Department of Geographical and Environmental Science, Queensgate, Huddersfield HD1 3DH.

R. Cummins	Institute of Terrestrial Ecology, Hill of Brathens, Banchory AB31 4BY.
F. Currie	The Forestry Authority, Great Eastern House, Tenison Rd., Cambridge CB1 2DU.
D.G. Dawson	The London Ecology Unit, 125 Camden High St., London NW1 7JR.
P. Dennis	MLURI, Craigiebuckler, Aberdeen AB9 2QJ.
P. Dimopoulos	Laboratory of Plant Ecology, Department of Biology, University of Patras, GR-26500, Greece
T. Docherty	British Trust for Ornithology, The Nunnery, Thetford, Norfolk IP24 2PU.
J.W.Dover	Myerscough College, Bilsborrow, Preston PR3 0RY.
J. Edmunds	Lancashire Wildlife Trust, Cuerden Valley Wildlife Centre, Shady Lane, Bamber Bridge, Preston PR5 6AU.
M.E. Evans	Wagtails, 8 Pennine View, Glasson Dock, Lancaster LA2 0AS.
S. Falk	Herbert Art Gallery & Museum, Jordan Well, Coventry CV1 5PQ.
A. Farina	Lunigiana Museum of Natural History and Laboratory of Landscape Ecology,54011 Aulla,Italy.
E. Ferron	School of Biology and Biochemistry, Queen's University Belfast, 97 Lisburn Rd., Belfast BT9 7BL.
L. Firbank	Institute of Terrestrial Ecology, Monks Wood, Abbots Ripton, Huntingdon PE17 2LS.

J. Ford	Department of Biological Sciences, Napier University, Colinton Rd., Edinburgh EH10 5DT.
G. Fry	Norwegian Institute for Nature Conservation, Urbygningen, P.O. Box 5064, The Norwegian Agricultural University, N-1432, Aas, Norway.
R. Gustavsson	Department of Landscape Planning, Box 58, S-230 53 Alnarp, Sweden.
D. Gwynne	38, Leven Rd., Troon, Ayrshire KA10 7DX.
S. Hinsley	Institute of Terrestrial Ecology, Monks Wood, Abbots Ripton, Huntingdon PE17 2LS.
R.J. Hobbs	CSIRO, Division of Wildlife & Ecology, LMB 4, Midland, Perth, Western Australia 6056
R.A. Hughes	Scottish Natural Heritage, 2, Anderson Place, Edinburgh EH6 5NP
A. Hull	Ponds Research Unit, Liverpool John Moores University, 15-21 Webster St., Liverpool L3 2ET.
D. Huntley	Myerscough College, Bilsborrow, Preston PR3 0RY.
M. Ihse	Department of Physical Geography, Stockholm University, S-10691 Stockholm, Sweden.
J. Jackson	Department of Environmental Biology, Nene College, Park Campus, Northampton NN2 7AL.
H. Kinnunen	Department of Zoology, University of Helsinki, P.O. Box 17, FIN-00014, University of Helsinki, Finland.
K. Kirby	English Nature, Northminster House, Peterborough PE1 1UA.
D. Knight	Leeds City Council, Countryside Service, Home Farm, Temple Newsam Estate, Leeds, LS15 0BJ.

D. McCollin	Department of Environmental Biology, Nene College, Park Campus, Northampton NN2 7AL.
D. McCracken	Joint Nature Conservation Committee, 25 High Calside, Paisley PA2 6BY.
G. Morries	Lancashire County Council. Planning Department, P.O. Box 160, East Cliff County Offices, Preston PR1 3EX.
M.C. North	Gillespies, Edmund House, Wood St., Altrincham, Cheshire WA14 1ED.
C.T. di Pablo	Department Ecologia, Fac. Biologia, Universidad Complutense, 28040 Madrid, Spain.
T. Pakkala	Department of Zoology, University of Helsinki, P.O. Box 17, FIN-00014, University of Helsinki, Finland.
S. Potter	Department of Planning and Economic Development, Staffordshire County Council, County Buildings, Martin St., Stafford ST16 2LE.
G. Price	The Forestry Authority, Great Eastern House, Tenison Rd., Cambridge CB1 2DU.
D. Ratcliffe	West Lancashire District Council, 52 Derby St., Ormskirk, Lancashire L39 2DF.
A. Reeves	Myerscough College, Bilsborrow, Preston PR3 0RY.
P.A. Robertson	The Game Conservancy, Fordingbridge, Hampshire SP6 1EF.
W. Robson	Norwegian Institute for Nature Conservation, Urbygningen, P.O. Box 5064, The Norwegian Agricultural University, N-1432, Aas, Norway.

M. Roe	Department of Town & Country Planning, Claremont Tower, University of Newcastle, Newcastle-upon-Tyne.
I. Sarlov-Herlin	Department of Landscape Planning, Box 58, S-230 53 Alnarp, Sweden.
A.G.M. Schotman	Instituut voor Bosen Natuuronderzoek, Bostandweg 20, Postbus 23, 6700\AA Wageningen, The Netherlands.
J. Slater	ADAS Cambridge, Brooklands Ave., Cambridge CB2 2BL.
T. Sparks	Institute of Terrestrial Ecology, Monks Wood, Abbots Ripton, Huntingdon PE17 2LS.
A. Stott	Department of the Environment, Room 921, Tollgate House, Houlton St., Bristol BS2 9DJ.
M.J.S.Swan	Amphibian Habitat Advisory Services, 19 St. Judith's Lane, Sawtry, Huntingdon PE17 5XE.
E.E. Thomas	Department of Environmental Planning & Management, Coleg Normal, Bangor, Gwynnedd, LL57 2PX.
J. Tiainen	Finnish Game & Fisheries Research Institute, P.O.Box 202, FIN-00151 Helsinki, Finland.
C. Topping	Scottish Agricultural College, Land Resources Dept., Doig Scott Building, Craibstone, Bucksburn, Aberdeen AB2 9TQ.
N.R. Webb	Institute of Terrestrial Ecology, Furzebrook Research Station, Wareham, Dorset BH20 5AS.
M. Wiren	Department of Landscape Planning, Box 58, S-230 53 Alnarp, Sweden.

I.P. Woiwod Entomology & Nematology Department, Rothamsted Experimental Station, Harpenden, Hertfordshire AL5 2JQ.

I.R. Wynne Entomology & Nematology Deparment, Rothamsted Experimental Station, Harpenden, Hertfordshire AL5 2JQ.

C. Yates 50, Chester St., South Freemantle 6162, Western Australia.

Preface

This volume comprises the written contributions to the third annual conference of IALE(UK) the UK region of the International Association for Landscape Ecology.

The UK region was founded at a meeting entitled 'Landscape Ecology in Britain' on 22-23rd September 1992 at Nottingham University. This followed two years during which a UK section of IALE International had been set up to encourage communication, interdisciplinary research, and the development of knowledge and interaction between scientists and those concerned with planning and management of the landscape in the UK.

We define LANDSCAPE ECOLOGY as the study of the interactions between the temporal and spatial aspects of a landscape and its flora, fauna and cultural components.

Although the conference on Fragmentation in Agricultural Landscapes was conceived and organised by the UK Region, the papers presented at the meeting reflect the international nature of the Association, with both speakers and delegates from different countries and even continents. IALE(UK) are indebted to all those who have given of their time and energy as platform speakers and as contributors to this volume.

PART I:

LANDSCAPES AND BIOTOPES

Fragmentation in the wheatbelt of Western Australia: landscape-scale problems and solutions

R.J. Hobbs

CSIRO, Division of Wildlife and Ecology, LMB 4, PO Midland, Western Australia 6056, Australia

Abstract

The wheatbelt of Western Australia has been extensively cleared for agriculture over the last century, resulting in the fragmentation of native vegetation into small remnant areas. These remnants are under threat from altered landscape processes, and will not, by themselves, maintain the existing biota. The nature of current threats is examined, and possible approaches to dealing with them are discussed. Long-term persistence of native ecosystems will probably be possible only if large areas are returned to natural or semi-natural vegetation. This approach requires the development of adequate methods of landscape planning and restoration, and the ability and willingness of the farming community to implement these. I discuss current research in these areas and assess the likelihood of success.

Introduction

Fragmentation of natural ecosystems is generally seen to be one of the most important threats to biodiversity worldwide. Fragmentation occurs when human activities such as agricultural development, extraction forestry or urbanization remove large proportions of the natural ecosystem and replace them with a greatly modified matrix, within which small remnants of the native ecosystems remain. This process results not only in a vastly reduced area of the natural ecosystems but also their subdivision into small and relatively isolated fragments.

The response of biota to ecosystem fragmentation has received much study in the conservation literature. This has largely centered on species and community responses to changes in ecosystem size and isolation, within the theoretical framework of island biogeography. Considerable debate has focused on such questions as whether a small number of large reserves will maintain native biota better than a larger number of smaller reserves, and whether reserves linked by corridors are better than those without linkages. Much of this debate has carried on with a paucity of data from real systems, and Saunders *et al.* (1991) have suggested **that it has contributed relatively little of practical value to nature conservation.** They pointed out that, until recently, relatively little emphasis had been placed on understanding the effects of fragmentation on the structural and functional aspects of natural and managed ecosystems.

It is from this perspective that I discuss the work on fragmentation that has been carried out in the Western Australian wheatbelt. I firstly discuss the history of fragmentation in the area, and examine the biogeographic work that was carried out in the 1970s. I then examine the ecosystem consequences of fragmentation, and point out the importance of the landscape context. Current attempts to develop integrated solutions to multiple problems at the landscape scale are discussed. Due to space limitations, I refer mostly to Western Australian work, but a large body of literature from other parts of the world which indicates that the Western Australian work is relevant to fragmented systems everywhere.

The Western Australian wheatbelt

The southwest of Western Australia is an ancient weathered landscape and has a diverse flora with a high degree of endemism (Hopper, 1979; Hopper, 1992; McArthur, 1993). The area was settled by Europeans during the first half of the 19th Century, but rapid development commenced only during this century. The area suitable for agriculture is determined by the distribution of suitable soils on one hand, and adequate rainfall on the other, and covers approximately 140,000 km^2 (Fig. 1).

Figure 1. Southwestern Australia, showing the wheatbelt and adjacent forest and uncleared areas.

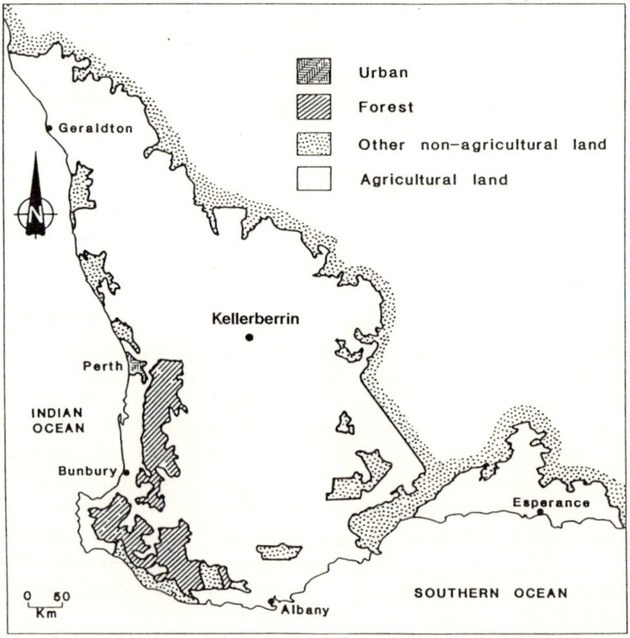

The area was extensively cleared of native vegetation during this century, with a rapid rise in the rate of clearing after 1945 (Saunders *et al.*, 1985). Over the wheatbelt as a whole, approximately 7% of native vegetation remains, although this percentage varies greatly within the region, with the oldest settled areas retaining less vegetation. Prior to clearing, the area consisted of a mosaic of vegetation types including various woodland, shrubland and mallee (multi-stemmed eucalypt) types. These vegetation types were cleared to differing extents, depending on the suitability of the underlying soils for agriculture. Woodlands, in particular, were heavily cleared, with some types being all but eliminated from the region (Beard and Sprenger, 1984).

Figure 2. A section of the central wheatbelt in Western Australia, to the north of the town of Kellerberrin, showing remnant vegetation, nature reserves and road verge vegetation (from Hobbs *et al.* 1993a).

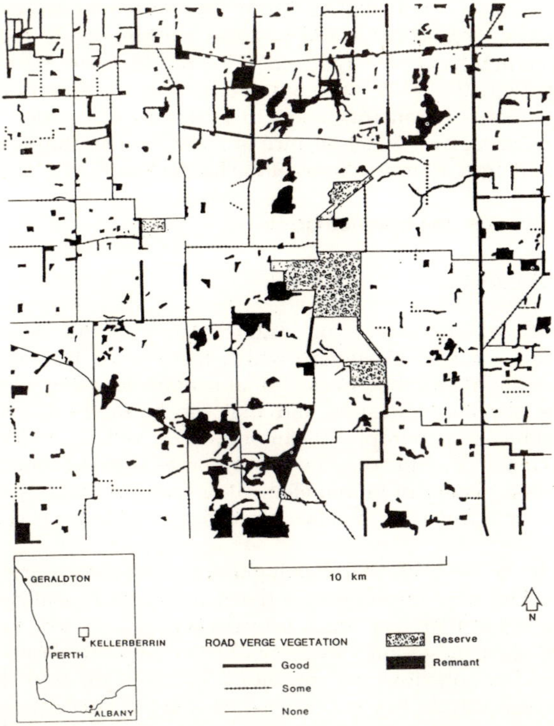

The native vegetation left in the region is present as a large number of mostly very small remnants. A few large areas are reserved as national parks or nature reserves, and these account for a large proportion of the native vegetation remaining (Wallace and Moore, 1987). Within the area administered by the Department of Conservation and Land Management as the wheatbelt region, there are over 600 nature reserves, which range in size from 0.4 ha to 309,000 ha. The three largest reserves account for 64% of the reserved area, and the median size of reserves is 114 ha (Wallace & Moore, 1987). In addition to the areas designated as nature reserves or set aside for other reasons (e.g. water reserves, townsite reserves) there are a great many remnants on private land. In some areas, these privately owned remnants constitute a large proportion of the remaining native vegetation. For instance, in a 1680 km^2 area studied in detail in the central wheatbelt, 77% of remnant vegetation is on private land, and is mostly in small patches, often less than 10 ha (Arnold and Weeldenburg, 1991; Saunders *et al.*, 1993a). In addition to this, vegetation has been retained along road verges, which often contain significant proportions of the vegetation remaining in the area. Road verges vary in width and condition and provide some degree of linkage between remnant areas (Figure 2; Hobbs *et al.*, 1993a).

The wheatbelt landscape has thus been rapidly transformed from a continuous mosaic of predominantly perennial vegetation types to a landscape dominated by agriculture based on annual crops and pastures, within which small fragments of the original vegetation have been retained. This transformation has had profound implications for the native biota, through habitat loss and isolation. Importantly, however, the process of transformation to agriculture has also fundamentally altered landscape processes, which in turn are now having important impacts on both the native and agricultural sections of the landscape.

Consequences of landscape fragmentation

Focusing on the fragments

Initial research on the effects of fragmentation in the Western Australian wheatbelt mirrored studies elsewhere which focused on the effects of size and isolation on the biota in remnant areas. Initial propositions suggested that remnant areas should be similar to islands and that the biota should follow the rules developed by island biogeography (Diamond, 1975; Wilson and Willis, 1975). A survey of a selected subsample of wheatbelt reserves showed that this was broadly the case, with the diversity of faunal elements depending largely on remnant size and, for some groups, diversity of vegetation types (Kitchener *et al.*, 1980a,b, 1982).

This approach was valuable in indicating that the remaining large reserves in the wheatbelt were extremely important for the retention of faunal diversity, but the studies also concluded that very much larger reserves would be needed to ensure the maintenance of existing biotic diversity. The largely correlative approach used in these studies did not, however, provide much insight into the mechanisms behind the observed trends, beyond suggesting that the effect of remnant size was probably linked to its effect on habitat diversity. Considerable variations in diversity within faunal groups could be observed in remnants of similar size and time since isolation, which could not necessarily be explained on the basis of biogeographic factors. There was also significant confounding of variables, with the reserves which were isolated earliest also being the smallest. This is a recurring problem when attempting to assess the relative importance of different factors likely to affect the biota within wheatbelt remnants.

Further studies of the biota within remnant areas have indicated that the fauna is still in a state of flux in the wheatbelt landscape. The wheatbelt has recorded the highest rates of decline and extinction of mammal species within Western Australia (Burbidge and McKenzie, 1989), and large changes in the distribution and abundance of bird species have occurred, and are still occurring (Saunders, 1989,1990,1993); Figure 3. The wheatbelt also has the highest numbers of threatened plant species of any area in Australia outside tropical Queensland (Briggs and Leigh, 1988; Hopper *et al.*, 1990).

Figure 3. The distribution of Carnaby's cockatoo (*Calyptorhynchus funereus latirostris*) in southwestern Australia in 1968 (left), based on information provided by schools, and in 1987 (right) based on sitings by volunteer observers (wheatbelt only). Closed symbols are recorded sightings, open circles are locations with no sightings (from Saunders 1990).

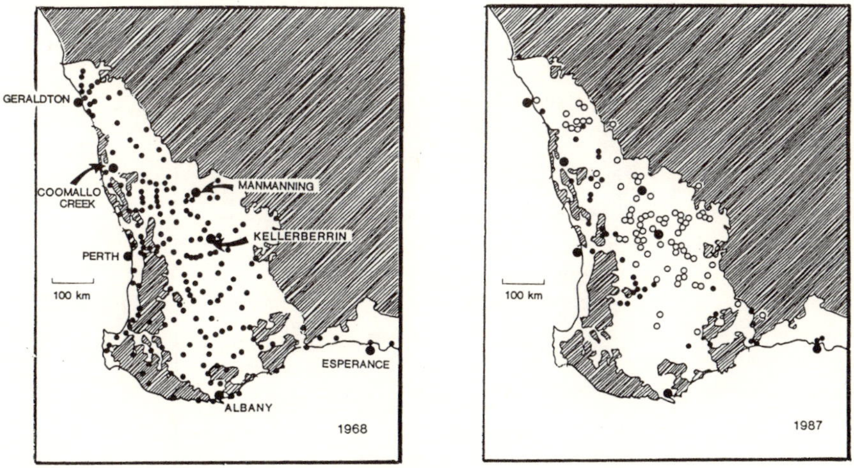

Focusing on the matrix

While the biogeographic processes operating as a result of habitat reduction and isolation undoubtedly have had, and continue to have, a profound impact on the biota of the wheatbelt, the alteration in landscape processes resulting from fragmentation has as large, if not larger impacts. Most studies of fragmented systems concentrate on the fragments, and most conservation management is conducted within individual fragments. Consider, for instance, the practice of drawing up management plans for individual nature reserves. This ignores the fact that the fragments of native vegetation take up only a small percentage of the landscape and are heavily influenced by the surrounding matrix of agricultural (or other production-based) land (Fig. 4).

The switch from continuous cover of native vegetation to small fragments in a predominantly altered matrix brings with it many important consequences for the remnant vegetation, over and above the biogeographic effects of reduced area and increased isolation. These have been reviewed by Saunders *et al.* (1991), and stem from changes in physical, chemical and biotic flows in the altered landscape. Also relevant are changes in the disturbance regime and the spread of disturbance through the fragmented landscape (Hobbs, 1987; Baker, 1992), with resulting changes in ecosystem dynamics (e.g. (Yates *et al.*, 1994).

Figure 4. A comparison of the traditional focus of research and management in fragmented landscapes on individual fragments (top) and the landscape approach, focusing on the matrix surrounding the fragments and the influence of factors arising from the matrix (bottom).

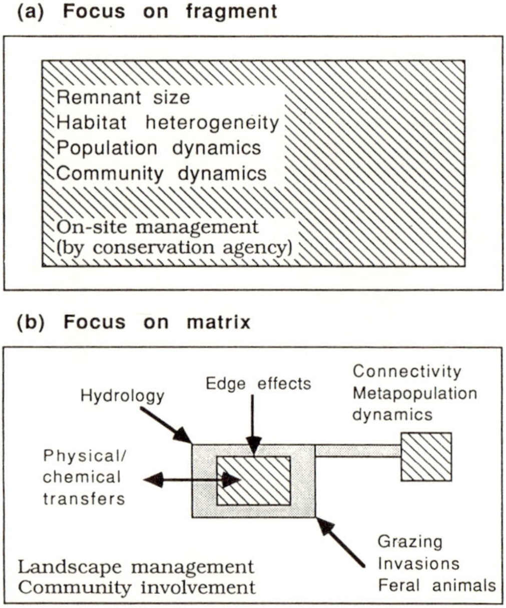

Changes in landscape fluxes

Hobbs (1993a) has discussed the changes in ecosystem processes which have occurred in the wheatbelt landscape. Profound changes in nutrient, energy and water fluxes result from the replacement of perennial vegetation with predominantly annual crops and pastures.

In terms of nutrients, the landscape changes from one in which considerable internal nutrient redistribution occurs, but little net loss is evident, to one in which large inputs of fertilizer occur, and large losses of nutrients result from wind and water erosion and harvesting (Hobbs 1993a). In the fragmented wheatbelt landscape, remnant native vegetation on low nutrient soils are set in an agricultural matrix with elevated nutrient levels and increased mobility of topsoil. Loss of topsoil by wind and water erosion , especially during episodic storm events, is a major problem over much of the wheatbelt and can cause significant long-term yield reductions (Nulsen, 1993). The impacts of this soil movement on remnant vegetation may also be profound, since water movement and wind bring nutrient-rich soil into remnant areas and hence alter the nutrient regime. This has important follow-on implications in terms of weed invasion, which is promoted by nutrient increase (Hobbs and Atkins, 1988).

Changed energy fluxes and wind regimes result from the change from tall evergreen vegetation to the alternating short annual vegetation and bare soil associated with agriculture (Hobbs 1993a). As well as influencing local microclimates, these changes may have profound regional scale implications. Rainfall patterns in southwestern Australia have changed over the past century, such that annual amounts have declined in the agricultural areas, but increased in the adjacent inland areas (Pittock, 1988). Smith *et al.*, (1992) have recently suggested that these changes could be associated with the effects of changing albedo on cloud formation. Evidence for this comes from a phenomenon frequently observed in satellite images for the area whereby clouds are found only above the large areas of native vegetation remaining in the wheatbelt and to the east of the line between cleared and uncleared land.

Perhaps the most important alteration to the wheatbelt landscape resulting from the extensive fragmentation of the native vegetation has been the effect on hydrology. In a mediterranean landscape with a pronounced summer drought, water availability is an important determinant of plant and animal distribution, growth and survival. Within uncleared vegetation, plant architecture, soil structure and soil biotic activity act to channel and redistribute rainfall, and most is utilized and transpired or evaporated, with little runoff or transfer to the water table (Nulsen *et al.*, 1986; McFarlane *et al.*, 1992,1993). Removal of native vegetation and replacement with annual crops and pastures results in considerably less efficient use of rainfall, increased runoff and greater inputs to the water table. This has led to rising water tables, with rises of as much as 50cm yr^{-1} recorded in some areas (George *et al.*, 1995). Beneath much of the wheatbelt are considerable quantities of salt stored at depth, resulting from a long history of input in ocean-derived rainfall coupled with poor drainage (Hingston and Galaitis, 1976). As water tables rise, the stored salt is mobilized, and wherever the water table reaches the surface, salinization occurs. Salinization and waterlogging have resulted in the loss of large areas of productive farmland, and pose a continuing threat to a significant proportion of the wheatbelt (Nulsen 1993; McFarlane *et al.* 1993).

Rising water tables also threaten the fragments of native vegetation. While native vegetation within remnants still uses water more efficiently than the surrounding crops and pastures, the overriding influence of the agricultural matrix means that water tables are rising even under large remnants (Salama *et al.*, 1994). Many low lying areas have already been affected, and most of the previously-fresh water lakes have gone saline (Froend *et al.*, 1987; Froend and McComb, 1991; Hobbs *et al.*, 1993b). In some areas the water table is rising so quickly that remnant vegetation will be destroyed within the next 5-10 years (George, *et al.* 1995), while in lower rainfall areas, the timescale is longer but the problem is no less insidious.

Such considerations are also relevant elsewhere, including the United Kingdom. The importance of surrounding land uses in altering the hydrological balance and hence compromising the conservation values of reserve areas is being increasingly recognized (Rowell, 1986; Green and Robins, 1993; Harding, 1993).

Biotic flows: from matrix to fragments

Domestic stock, predominantly sheep, affect remnant areas in several important ways by removing understorey vegetation, preventing regeneration of dominant species and altering soil structure (Scougall *et al.*, 1993; Norton *et al.*, 1994; Norton & Hobbs, in prep.). Movement of domestic stock into remnants significantly increases nutrient input, since stock feed in the paddocks and defecate in the remnants (Scougall *et al.*, 1993). They thus represent an important influence on remnants which arises from the surrounding matrix.

Another particularly important influence on the native biota in remnants is the influx of feral and invading animals. Foxes (*Vulpes vulpes*) , cats (*Felis catus*) and rabbits (*Oryctolagus cuniculus*) are particularly important in the context of the Western Australian wheatbelt. Although rabbits were introduced to eastern Australia in the middle of the last century, they reached southwestern Australia only in the early part of this century (Stodart and Parer, 1988). Fox and cat invasions probably occurred at about the same time. Rabbits reached plague proportions and led to large-scale habitat modification, while foxes and cats predated heavily on native mammals, ground- and hollow-nesting birds, lizards and insects (Hobbs *et al.* 1993b). While most emphasis has been placed on fragmentation as the cause of declines in faunal populations, the commencement of rapid fragmentation coincided with the arrival of rabbits and foxes. It is likely that these invading species, and the measures used to control them by the human population, were responsible for much of the faunal (especially mammal) decline (Hobbs *et al.* 1993b). This hypothesis is supported by recent work which has sought to reduce fox numbers in certain areas. Such reductions have resulted in significant increases in the populations of such native mammals as the black footed rock wallaby (*Petrogale lateralis*), numbat (*Myrmecobius fasciatus*) and bilby (*Macrotis lagotis*), either where populations of these species remained, or where they have been reintroduced (Kinnear *et al.*, 1988; Friend, 1990).

As well as the species introduced by humans, other species have spread into the wheatbelt as a result of the extensive habitat modification. Particularly important are the arid zone species such as the galah (*Cacatua rosiecapilla*), which has spread throughout the wheatbelt region and competes with other cockatoos for nest hollows and damages or sometimes kills trees by ringbarking them (Saunders, 1993).

Non-native plant species also represent a significant biotic flow into remnant areas from the surrounding matrix. The native vegetation is mostly composed of long-lived perennial species with many adaptations to low nutrient availability. Remnants are now surrounded by a matrix of annual species, most of which will not establish or grow well in the unmodified soils within remnants. However, if the soil within remnants is disturbed or enriched, the annual species can establish (Hobbs and Atkins, 1988; Hobbs, 1991). Some vegetation types, especially shrublands, are remarkably resistant to weed invasion, whereas others, such as woodlands are more susceptible. Weeds are often a problem only at the edge of remnants (Hester and Hobbs, 1992), but stock grazing can allow weed invasion throughout entire remnants (Scougall *et al..*, 1993). The intactness of the vegetation cover and the degree of soil disturbance are primary determinants of the degree to which weeds can invade. A large proportion of remnants within the Kellerberrin study area are invaded to some extent (Saunders *et al.*, 1993a). Once weeds have invaded an area, it is very difficult to remove them, and they prevent regeneration of perennial species (Hobbs and Atkins, 1991). Plant invasions thus represent one of the most important threats to the maintenance of native vegetation in remnant areas.

Biotic flows: connectivity

As well as biotic flows from the matrix into remnant areas, flows of biota between remnants through the matrix need to be considered. Most remnant areas are likely to be too small to support populations of many native animal species, and hence these need to move around the landscape to fulfill their resource and habitat requirements. The extent to which this movement occurs has been the subject of intensive research in the Kellerberrin area. Of particular interest has been the importance of road verge vegetation as corridors or conduits between remnants. Much debate has centered on the utility of corridors, following the initial enunciation of their value in terms of biogeographic design (Noss, 1987; Simberloff and Cox, 1987; Harris and Scheck, 1991; Hobbs, 1992; Simberloff *et al.*, 1992). Studies of birds and kangaroos in the Kellerberrin area have indicated that considerable movement between remnants does occur, at least for some species (Saunders and de Rebeira, 1991; Arnold *et al.*, 1993). These authors also suggest that corridors facilitate movement by some species, even if movement can occur in the absence of corridors. On the other hand, Norton *et al.* (1994) presented evidence which suggested that corridors were not used in the same way as remnant areas by mistletoe-dispersing birds. Hobbs (1992) suggested that unequivocal data on the requirement of corridors for movement would be very difficult to obtain. Recent radiotelemetry studies by (Lynch *et al.*, 1995) have indicted that the movement function of corridors may actually be secondary to an important habitat function.

Corridors are therefore an important component of the conservation network in the wheatbelt. This importance is amplified by the fact that road verges often contain significant proportions of the total area of certain vegetation types within an area (Arnold and Weeldenburg, 1991), and retain many populations of rare flora (Hopper *et al.*, 1990; Hobbs *et al.*, 1993a). Corridors are, however, more prone to the many external (i.e. matrix-derived) influences discussed above than remnant areas, since their linearity leads to a high edge: area ratio. Our studies have suggested that they are more susceptible to nutrient enrichment, weed invasion and episodic insect attack than equivalent remnant areas (Cale and Hobbs, 1991; Hobbs and Atkins, 1991; van Schagen *et al.*, 1992).

Can we understand the dynamics of fragmented systems?

The above discussion illustrates that ecosystem fragmentation does not entail a simple reduction in area and increase in isolation. Accompanying these changes is a wide array of other impacts. Studies of the dynamics of wheatbelt remnants have led me to conclude that it may be very difficult to understand these dynamics since we cannot differentiate between the effects of fragmentation *per se* versus the effects of matrix-derived influences. I have already discussed the relative roles of fragmentation versus fox predation in the decline of wheatbelt mammals. As another example, Norton *et al.* (1994) studied the distribution of mistletoes in woodland remnants, but found that the influence of remnant size was strongly confounded with the influence of stock grazing. Similarly, Yates *et al.* (1994) were able to understand the regeneration dynamics of an important wheatbelt woodland species, *Eucalyptus salmonophloia*, only by studying populations in extensive unfragmented areas to the east of the wheatbelt. They concluded that, without these adjacent unfragmented areas, little insight into how to manage fragmented populations would have been possible.

The difficulty of interpreting *post hoc* what is happening in fragmented systems is also highlighted by recent studies in Mexico, where studies of recently fragmented forest areas have shown that regeneration dynamics and vegetation structure are markedly altered, not by fragmentation *per se*, but by the loss of large herbivores (Dirzo and Miranda, 1990,1991; Dirzo, 1992). Without some knowledge of the dynamics of the unfragmented system, such insights would have been impossible. This therefore suggests that we should be cautious in our attempts to understand long-fragmented systems. What we observe may make little sense, simply because the dynamics are now far removed from those prevailing in the original system. This also influences the way we should go about managing these systems.

Finding management solutions

The catalogue of threats to remnant vegetation and the entire wheatbelt landscape presented above may appear to present an insurmountable challenge. However, the importance of the unique biota retained in the remnants and the importance of the agricultural enterprise in the area demand that solutions to the problems be found. I and others have argued elsewhere that the problems faced by conservation and agriculture stem from the same basic cause - i.e. the extensive removal of the native vegetation - and hence common solutions to the problems should be available (Hobbs and Saunders, 1991; Hobbs *et al.*, 1993a). From an agricultural viewpoint, measures are needed which reverse the current trends of degradation and loss of productive land. A variety of approaches to this are possible, but all involve replacing the current annual-based agriculture with a more diverse system which includes perennial and deep-rooted pastures and extensive revegetation (Lefroy *et al.*, 1992, 1993). From a conservation viewpoint, the requirement is for a reduction in the deleterious effects of matrix-derived influences on remnants and an increase in the area and connectivity of habitat.

There is currently a major push by government agencies and an increasing interest from the farming community for revegetation programs to reduce land degradation. While it seems possible that such revegetation could also be beneficial in a nature conservation sense, there is little evidence that this would, in fact, be the case, and only basic guidelines are available as to how revegetation could best be planned and implemented to benefit conservation (Hobbs, 1993b; Recher, 1993; Saunders *et al.*, 1993b). To be of conservation value, revegetation would need to be used to enhance the existing system of remnants. Revegetation could be used to provide buffer zones around remnant areas, corridors between remnants, or as additional habitat (Fig. 5; Hobbs 1993b). Nevertheless, conservation *per se* will rarely be the primary driver of revegetation efforts, and has therefore to be integrated with activities designed with the broader agricultural perspective in mind. An integrated approach to the planning and design of revegetation is therefore essential (Saunders, 1994).

Figure 5. Revegetation can be used to enhance the existing fragmented system by providing (a) buffer zones to protect existing remnants from edge effects, (b) corridors to increase connectivity between remnants, or (c) additional habitat to increase the effective size of a remnant (from Hobbs 1993b).

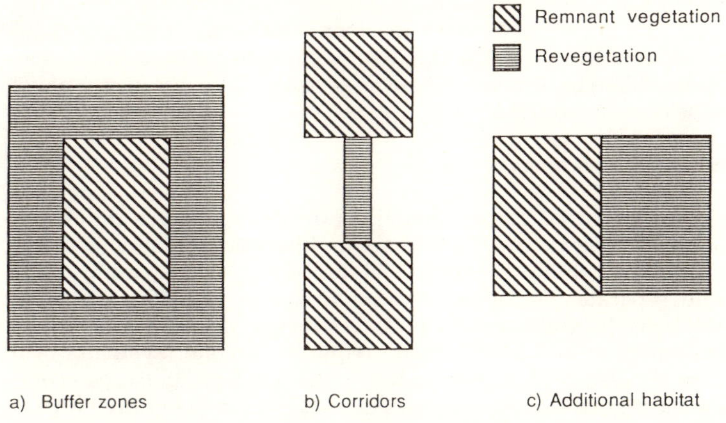

All of the above options are a distinct departure from traditional conservation management which has focused almost entirely within individual remnant/reserve areas, and has been conducted by conservation agencies largely in isolation from (or in opposition to) surrounding land managers. The suggested approach requires a broader perspective which encompasses the entire landscape, including the conservation and production components. It also requires increased cooperation between neighbouring land owners and managers. Conservation agencies which have previously often ignored the human community surrounding nature reserves are now becoming dependent on the cooperation and goodwill of that community to ensure the persistence of the reserves and of the non-reserved remnants which form an integral part of the conservation network. This will be achieved only if the local community recognizes the value of retaining remnant vegetation and agrees with the objectives of conservation management. Since the local community will also carry out much of the planning and management, it is essential that it is well informed, well motivated, and has access to the expert assistance and guidance needed for it to develop programs which tackle problems in an integrated and efficient way.

There is thus a strong social component to the management of fragmented landscapes, the importance of which cannot be underestimated. In the Western Australian wheatbelt, under-resourcing of the main conservation agency means that it cannot effectively manage the areas under its jurisdiction (Wallace and Moore, 1987): i.e. it cannot undertake the necessary fragment-focused management, far less the broader matrix-focused issues. The persistence of the native biota over much of the wheatbelt thus depends on the activities of local communities. The development of community Land Care Groups and catchment groups (Goss and Chatfield, 1993) illustrates a growing awareness of the need for a landscape-scale approach to the problems facing land managers. This approach is not easy, since it involves cooperation, discussion, compromise and an additional layer of responsibility over and above the traditional responsibility for one's own land. The successful reversal of current trends of degradation of both conservation and agricultural values depends on the successful development of techniques for integrated landscape management. This in turn depends on the success of attempts to bring local communities and all concerned parties together to tackle the problems. The trends are encouraging, but there is still a long way to go, with many scientific, social and economic issues still to be resolved. These three sets of issues are inextricably linked, and have to be resolved together.

The role of landscape ecology

The need for increased cooperation also extends to research into fragmented landscapes. The problems encountered are multifaceted, and no one discipline is sufficient on its own to tackle them. The increasing fragmentation of scientific research into separate disciplines and subdiscipines militates against finding adequate solutions. Instead interaction between diverse disciplines such as ecology, hydrology, economics and sociology are required. Greater synthesis of diverse approaches is also required. As an example, the fields of landscape ecology and conservation biology have many common threads but have interacted little (Hobbs, 1994). The landscape is the relevant scale at which many of today's environmental problems have to be tackled, and yet landscape ecologists have made relatively little effort to develop approaches to these problems. Landscape ecologists have a responsibility to take up the challenge of developing landscape ecology into a science which can usefully contribute to the cooperative effort which is needed as a response to the problems posed by fragmentation and other types of environmental degradation.

References

Arnold, G.W.; Steven, D.E.; Weeldenburg, J.R. & Smith, E.A. (1993) Influences of remnant size, spacing pattern and connectivity on population boundaries and demography in euros *Macropus robustus* living in a fragmented landscape. *Biological Conservation* **64**: 219-230.

Arnold, G.W. & Weeldenburg, J.R. (1991) *The distributions and characteristics of remnant native vegetation in parts of the Kellerberrin-Tammin-Trayning and Wyalkatchem Shires of Western Australia.* CSIRO, Perth.

Baker, W.L. (1992) The landscape ecology of large disturbances in the design and management of nature reserves. *Landscape Ecology* **7**: 181-194.

Beard, J.S. & Sprenger, B.S. (1984) *Geographical data from the vegetation survey of Western Australia, Vegetation Survey of Western Australia.* Vegmap Publications, Applecross, Western Australia.

Briggs, J.D. & Leigh, J.H. (1988) *Rare or Threatened Australian Plants (1988 Revised Edition). Special Publication 14.* Australian National Parks and Wildlife Service, Canberra.

Burbidge, A.A. & McKenzie, N.L. (1989) Patterns in the modern decline of Western Australia's vertebrate fauna: causes and conservation implications. *Biological Conservation* **50**: 143-198.

Cale, P. & Hobbs, R.J. (1991) Condition of roadside vegetation in relation to nutrient status. D.A. Saunders. and R.J. Hobbs. (Eds.). *Nature Conservation 2: The role of corridors,* Surrey Beatty and Sons, Chipping Norton, NSW 363-362

Diamond, J.M. (1975) The island dilemma: lessons of modern biogeographic studies for the design of nature reserves. *Biological Conservation* **7**: 129-46.

Dirzo, R. (1992) Rates of deforestation in Los Tuxtlas, a neotropical area in southeast Mexico. *Conservation Biology* **6**: 84-90.

Dirzo, R. & Miranda, A. (1990) Contemporary neotropical defaunation and forest structure, function and diversity - a sequel to John Terborgh. *Conservation Biology* **4**: 444-447.

Dirzo, R. & Miranda, A. (1991) Altered patterns of herbivory and diversity in the forest understory: a case study of the possible consequences of contemporary defaunation. P.W. Price, T.M. Lewinsohn, G.W. Fernandes and W.W. Benson (Eds). *Plant-Animal Interactions: Evolutionary Ecology in Tropical and Temperate Regions,* Wiley and Sons, Chichester

Friend, J.A. (1990) The numbat *Myrmecobius fasciatus* (Myrmecobiidae): History of decline and potential for recovery. *Proceedings of the Ecological Society of Australia* **16**: 369-377.

Froend, R.H.; Heddle, E.M.; Bell, D.T. & McComb, A.J. (1987) Effects of salinity and waterlogging on the vegetaion of Lake Toolibin, Western Australia. *Australian Journal of Ecology* **12**: 281-298.

Froend, R.H. & McComb, A.J. (1991) An account of the decline of Lake Towerrinning, a wheatbelt wetland. *Journal of the Royal Society of Western Australia* **73**: 123-128.

George, R.J.; McFarlane, D.J. & Speed, R.J. (1995) The consequences of a changing hydrologic environment for native vegetation in south Western Australia. D.A. Saunders, J.L. Craig and E.M. Mattiske (Eds.). *Nature Conservation 4: the role of networks,* Surrey Beatty and Sons, Chipping Norton, NSW in press

Goss, K.F. & Chatfield, J. (1993) Landcare groups in Western Australia: the role of self help groups in restoring degraded farmland. D.A. Saunders, R.J. Hobbs and P.R. Ehrlich (Eds.). *Nature Conservation 3: reconstruction of fragmented ecosystems, global and regional perspectives,* Surrey Beatty, Chipping Norton, NSW 281-293

Green, R.E. & Robins, M. (1993) The decline of the ornithological importance of the Somerset Levels and moors, England and changes in the management of water levels. *Biological Conservation* **66**: 95-106.

Harding, M. (1993) Redgrave and Lopham Fens, East Anglia, England: a case study of change in flora and fauna due to groundwater abstraction. *Biological Conservation* **66**: 35-45.

Harris, L.D. & Scheck, J. (1991) From implications to application: the disperal corridor principle applied to conservation of biological diversity. D.A. Saunders and R.J. Hobbs (Eds.). *Nature Conservation 2: the role of corridors,* Surrey Beatty and Sons, Chipping Norton, Australia 189-220

Hester, A.J. & Hobbs, R.J. (1992) Influence of fire and soil nutrients on native and non-native annuals at remnant vegetation edges in the Western Australian wheatbelt. *Journal of Vegetation Science* **3**: 101-108.

Hingston, F.J. & Galaitis, V. (1976) The geographic variation of salt precipitated over Western Australia. *Australian Journal of Soil Research* **14**: 319-335.

Hobbs, R.J. (1987) Disturbance regimes in remnants of natural vegetation. D.A. Saunders, G.W. Arnold, A.A. Burbidge and A.J.M. Hopkins (Eds.). *Nature Conservation: the role of remnants of native vegetation,* Surrey Beatty and Sons, Chipping Norton, NSW 233-240

Hobbs, R.J. (1991) Disturbance as a precursor to weed invasion in native vegetation. *Plant Protection Quarterly* **6**: 99-104.

Hobbs, R.J. (1992) Corridors for conservation: solution or bandwagon? *Trends in Ecology and Evolution* **7**: 389-392.

Hobbs, R.J. (1993a) Effects of landscape fragmentation on ecosystem processes in the Western Australian wheatbelt. *Biological Conservation* **64**: 193-201.

Hobbs, R.J. (1993b) Can revegetation assist in the conservation of biodiversity in agricultural areas? *Pacific Conservation Biology* **1**: 29-38.

Hobbs, R.J. (1994) Landscape ecology and conservation: moving from description to application. *Pacific Conservation Biology* **1**: in press.

Hobbs, R.J. & Atkins, L. (1988) Effect of disturbance and nutrient addition on native and introduced annuals in the Western Australian wheatbelt. *Australian Journal of Ecology* **13**: 171-179.

Hobbs, R.J. & Atkins, L. (1991) Interactions between annuals and woody perennials in a Western Australian wheatbelt reserve. *Journal of Vegetation Science* **2**: 643-654.

Hobbs, R.J. & Saunders, D.A. (1991) Reintergrating fragmented landscapes: a preliminary framework for the Western Australian wheatbelt. *Journal of Environmental Management* **33**: 161-167.

Hobbs, R.J.; Saunders, D.A. & Arnold, G.W. (1993a) Integrated landscape ecology: a Western Australian perspective. *Biological Conservation* **64**: 231-238.

Hobbs, R.J.; Saunders, D.A.; Lobry de Bruyn, L.A. & Main, A.R. (1993b) Changes in biota. R.J. Hobbs and D.A. Saunders (Eds). *Reintegrating Fragmented Landscapes. Towards sustainable production and nature conservation,* Springer-Verlag, New York 65-106

Hopper, S.D. (1979) Biogeographical aspects of speciation in the southwest Australian flora. *Annual Review of Ecology and Systematics* **10**: 399-422.

Hopper, S.D. (1992) Patterns of plant diversity at the population and species level in south-west Australian mediterranean ecosystems. R.J. Hobbs (Ed). *Biodiversity of Mediterranean Ecosystems in Australia,* Surrey Beatty and Sons, Chipping Norton, NSW 27-46

Hopper, S.D.; van Leeuwen, S.; Brown, A.P. & Patrick, S.J. (1990) *Western Australia's Endangered Flora and Other Plants Under Consideration for Declaration.* Department of Conservation and Land Management, Perth.

Kinnear, J.E.; Onus, M.L. & Bromilow, R.N. (1988) Fox control and rock wallaby population dynamics. *Australian Wildlife Research* **15**: 435-450.

Kitchener, D.J.; Chapman, A.; Dell, J.; Muir, B.G. & Palmer, M. (1980a) Conservation value for mammals of reserves in the Western Australian wheatbelt. *Biological Conservation* **18**: 179-207.

Kitchener, D.J.; Chapman, A.; Dell, J.; Muir, B.G. & Palmer, M. (1980b) Lizard assemblage and reserve size and structure in the Western Australian wheatbelt - some implications for conservation. *Biological Conservation* **17**: 25-62.

Kitchener, D.J.; Dell, J.; Muir, B.G. & Palmer, M. (1982) Birds in Western Australian wheatbelt reserves - implications for conservation. *Biological Conservation* **22**: 127-163.

Lefroy, E.C.; Hobbs, R.J. & Scheltema, M. (1993) Reconciling agriculture and nature conservation: toward a restoration strategy for the Western Australian wheatbelt. D.A. Saunders, R.J. Hobbs and P.R. Ehrlich (Eds). *Nature Conservation 3: reconstruction of fragmented ecosystmems , Ggobal and regional perspectives,* Surrey Beatty and Sons, Chipping Norton, NSW 243-227

Lefroy, E.C.; Salerian, J. & Hobbs, R.J. (1992) Intergrating economic and ecological considerations. A theoretical framework. R.J. Hobbs and D.A. Saunders (Eds). *Reintegrating Fragmented Landscapes. Towards Sustainable Production and Nature Conservation,* Springer-Verlag, New York 209-244

Lynch, J.F.; Carmen, W.F.; Saunders, D.A. & Cale, P. (1995) Short-term use of vegetated road verges and habitat patches by four bird species in the central wheatbelt of Western Australia. D.A. Saunders, J.L. Craig and E.M. Mattiske (Eds). *Nature Conservation 4: the role of networks,* Surrey Beatty and Sons, Chipping Norton, NSW in press

McArthur, W.M. (1993) History of landscape developement. R.J. Hobbs and D.A. Saunders (Eds.). *Reintegrating Fragmented Landscapes. Towards Sustainable Production and Nature Conservation,* Springer-Verlag, New York 10-22

McFarlane, D.J.; George, R.J. & Farrington, P. (1993) Changes in the hydrologic cycle. R.J. Hobbs and D.A. Saunders (Eds). *Reintergrating Fragmented Landscapes. Towards Sustainable Production and Nature Conservation,* Springer-Verlag, New York 146-86

McFarlane, D.J.; Howell, M.R.; Ryder, A.T. & Orr, G.J. (1992) The effect of agricultural development on the physical and hydraulic properties of four Western Australian soils. *Australian Journal of Soil Research* **30**: 517-532.

Norton, D.N.; Hobbs, R.J. & Atkins, L. (1994) Fragmentation, disturbance and plant distribution: mistletoes in woodland remnants in the Western Australian wheatbelt. *Conservation Biology* in press.

Noss, R.F. (1987) Corridors in real landscapes: A reply to Simberloff and Cox. *Conservation Biology* **1**: 159-164.

Nulsen, R.A. (1993) Changes in soil properties. R.J. Hobbs and D.A. Saunders (Eds). *Reintegrating Fragmented Landscapes. Towards Sustainable Production and Nature Conservation,* Springer-Verlag, New York 107-145

Nulsen, R.A.; Bligh, K.J.; Baxter, I.N.; Solin, E.J. & Imrie, D.H. (1986) The fate of rainfall in a mallee and heath vegetated catchment in southen Western Australia. *Australian Journal of Ecology* **11**: 361-371.

Pittock, A.B. (1988) Actual and anticipated changes in Australia's climate. G.I. Pearman (Ed). *Greenhouse: Planning for Climatic Change,* CSIRO, Melbourne 35-51

Recher, H.F. (1993) The loss of biodiversity and landscape restoration: conservation, management, survival - an Australian perspective. D.A. Saunders, R.J. Hobbs and P. Ehrlich (Eds). *Nature conservation 3: reconstruction of fragmented ecosystems, global and regional perspectives.* Surrey Beatty & Sons, Chipping Norton 141-151

Rowell, T.A. (1986) The history of drainage at Wicken Fen, Cambridgeshire, England, and its relevance to conservation. *Biological Conservation* **35**: 111-142.

Salama, R.B.; Bartle, G.A. & Farrington, P. (1994) Water use of plantation *Eucalyptus camaldulensis* estimated by groundwater hydrograph seperation techniques and heat pulse method. *Journal of Hydrology* **156**: 163-180.

Saunders, D.A. (1989) Changes in the avifauna of a region, district and remnant as a result of fragmentation of native vegetation: the wheatbelt of Western Australia. A case study. *Biological Conservation* **50**: 99-135.

Saunders, D.A. (1990) Problems of survival in an extensively cultivated landscape: the case of Carnaby's Cockatoo *Calyptorhynchus funereus latirostris. Biological Conservation* **54**: 277-290.

Saunders, D.A. (1993) Community based observer scheme to assess avian response to habitat reduction and fragmentation in south western Australia. *Biological Conservation* **64**: 203-218.

Saunders, D.A. (1994) Can we integrate nature conservation with agricultural production? *Landscape and Urban Planning* **28**: 63-71.

Saunders, D.A. & de Rebeira, C.P. (1991) Values of corridors to avian populations in a fragmented landscape. D.A. Saunders and R.J. Hobbs (Eds.). *Nature Conservation 2: the role of corridors,* Surrey Beatty, Chipping Norton 221-240

Saunders, D.A.; Hobbs, R.J. & Arnold, G.W. (1993a) The Kellerberrin project on fragmented landscapes: A review of current information. *Biological Conservation* **64**: 185-192.

Saunders, D.A.; Hobbs, R.J. & Ehrlich, P.R. (1993b) Reconstruction of fragmented ecosystems: problems and possibilities. D.A. Saunders, R.J. Hobbs and P.R. Ehrlich (Eds.). *Nature Conservation 3: reconstruction of fragmented ecosystems, global and regional perspectives,* Surrey Beatty and Sons, Chipping Norton, NSW 305-313

Saunders, D.A.; Hobbs, R.J. &Margules, C.R. (1991) Biological consequences of ecosystem fragmentation: a review. *Conservation Biology* **5**: 18-32.

Saunders, D.A.; Rowley, I. & Smith, G.T. (1985) The effects of clearing for agricutlure on the distibution of cockatoos in the southwest of Western Australia. A. Keast, H.F. recher, H. Ford and D.A. Saunders (Eds). *Birds of Eucalypt forests and Woodlands: ecology, conservation, management,* Surrey Beatty & Sons, Chipping Norton, NSW 309-321

Scougall, S.A.; Majer, J.D. & Hobbs, R.J. (1993) Edge effects in grazed and ungrazed Western Australian wheatbelt remnants in relation to ecosystem reconstruction. D.A. Saunders, R.J. Hobbs and P.R. Ehrlich (Eds). *Nature Conservation 3: reconstruction of fragmented ecosystems , global and regional perspectives,* Surrey Beatty and Sons, Chipping Norton, NSW 163-178

Simberloff, D. & Cox, J. (1987) Consequences and costs of conservation corridors. *Conservation Biology* **1**: 63-71.

Simberloff, D.; Farr, J.A.; Cox, J. & Mehlman, D.W. (1992) Movement corridors: conservation bargains or poor investments? *Conservation Biology* **6**: 493-504.

Smith, R.C.G.; Huang, X.; Lyons, T.J.; Hacker, J.H. & Hick, P. (1992) Change in land surface albedo and temperature in southwestern Australia following the repacement of native perennial vegetation: satellite observations. Paper No. IAF-92-0117. 43rd Congress of the International Astronautical Federation, Washington, D.C.

Stodart, E. & Parer, I. (1988) *Colonisation of Australia by the Rabbit, Orycltolagus cuniculatus (L.).* CSIRO, Division of Wildlife and Ecology, Canberra.

van Schagen, J.J.; Hobbs, R.J. & Majer, J.D. (1992) Defoliation of trees in roadside corridors and remnant vegetation in the Western Australian wheatbelt. *Journal of the Royal Society of Western Australia* **75**: 75-82.

Wallace, K.J. & Moore, S.A. (1987) Management of remnant bushland for nature conservation in agricultural areas of Southwestern Australia - operational and planning perspectives. D.A. Saunders, G.W. Arnold, A.A. Burbidge and A.J.M. Hopkins (Eds). *Nature Conservation: the role of remnants of native vegetation,* Surrey Beatty and Sons, Chipping Norton, NSW 259-268

Wilson, E.O. & Willis, E.O. (1975) Applied biogeography. M.L. Cody and J.M. Diamond (Eds.). *Ecology and Evolution of Communities,* Belknap Press, Cambridge, Mass. 522-34

Yates, C.J.; Hobbs, R.J. & Bell, R.W. (1994) Landscape-scale disturbances and regeneration of semi-arid woodlands of south-western Australia. *Pacific Conservation Biology* **1**: in press.

FRAGMENTATION IN AGRICULTURAL LANDSCAPES

The habitat, the biotope and the landscape

N R Webb

Furzebrook Research Station, NERC Institute of Terrestrial Ecology,
Wareham, Dorset, BH20 5AS, UK

Abstract

The biodiversity of fragments of semi-natural communities in the landscape have formed the basis of many previous studies in which the effects of area and isolation on species populations have been examined. In this paper, a new approach is described which distinguishes between the biotope (fragments of semi-natural or natural communities) and the habitat. The habitat patch, which occurs within the biotope patch, is seen as the correct scale for examining the responses of species. The pattern of occurrence of habitat patches can be examined in relation to the distribution of biotope fragments. An analysis using heathland fragments containing habitats for heathland species in Dorset (southern England) is outlined as an example.

Introduction

We are familiar with the concept that many of the semi-natural communities of lowland Britain exist only as scattered fragments in a highly-modified and intensively-managed landscape. These fragments contain species and communities which,because they are remnants of the former landscape,are highly valued for conservation. The fragments may be scattered over a wide area or be clustered, reflecting the distribution of suitable soils and topography. In a sense, these remnant patches can be considered secondary. Originally, the post-glacial landscape was a patchwork of different vegetation types, the pattern being determined by variations in soils and topography (environmental resource patches *sensu* Foreman & Godron 1986). Early human activity created a series of disturbance patches which enabled elements of the current vegetation to establish, often over wide areas such as downland or heathland, and traditional management maintained these patches. In recent times they have been whittled away to create the present series of secondary remnant patches. They are secondary since they are remnants of the disturbance patches (and hence are examples of semi-natural communities) and not of the original resource patches.

Resources to conserve these patches may be limited and a selection will need to be made. Further, once a selection has been made, how should we manage the patches to maintain their conservation value? What is the appropriate way for an ecologist to view such patches when he wishes to describe patterns of distribution and to examine the dynamics of both the species populations and the entire communities?

Defining Patches

The simplest approach is to mark on a topographical map the extent of the communities. These may be defined qualitatively or by mapping the extent of some defined unit, such as those generated by the National Vegetation Classification. Frequently, the boundaries of the patches are fixed and clearly marked because they represent patterns of ownership or management. Elsewhere, even at this scale, boundaries may be difficult to recognise as one vegetation type grades into another - for example, where does the boundary lie between wet heath and dry? Nevertheless, when questions such as how patches are defined have been settled, it is possible to produce a map showing the distribution of these patches in the landscape. Field surveys or the interpretation of air photography have been used in many instances to produce such maps.

For the Dorset heaths we have used another approach (Chapman, Clarke & Webb 1989). The cover of the principal heathland vegetation types (broadly similar to NVC categories but easier to recognise in the field) have been recorded for 3110 200x200m grid squares covering the entire Dorset heathlands. Adjacent squares were joined to form heathland patches if they touched along their sides at any value of cover for the heathland vegetation, or if they touched at the corner when one square had at least 75% cover of heathland vegetation. This process generated 141 patches of heathland, and this broadly matched the distribution of patches derived by conventional mapping procedures. Maps of this type represent the scale at which humans perceive the landscape and, in particular, it is the scale at which land is owned and managed (Webb & Thomas, 1994).

Large Scale Stability

There have been considerable losses of heathland (>85% of the area since the mid-18th century), the loss is now very small (Webb 1990) and the arrangement of the patches can be considered to be stable. This is evident not only in their size, shape and position but in their composition. The apparent stability of the heaths when viewed at this scale has been highlighted in recent work (Bullock & Webb in press). This showed that despite extensive fires on the Dorset heathlands in 1976, when considered over a period of ten years and over the entire extent of the heathlands, there was negligible change in the composition of the vegetation and almost no effect of these fires was detectable ten years after they occurred. The only long-term effect detected was in the composition of scrub, which had shifted slightly from the dominance of gorse (*Ulex europaeus*) to that of birch (*Betula* spp.). It should be noted that, over the ten-year period considered, there was almost no management on these heaths, and that good conservation management would be expected to increase stability.

Clearly, the stability reported can be explained by the wide spatial and temporal scales at which change was examined. This approach provided an overview of the responses in a dynamic mosaic of heathland vegetation over a large area rather than the response in a small part of the mosaic. Examining the heaths at a smaller spatial or a shorter temporal scale would have revealed local extinctions of species which would have been unwelcome to conservationists.

Patterns of biodiversity

A patchwork of fragments defined in the way outlined above can be used to examine patterns of biodiversity, which in turn might be helpful in deciding conservation priorities. An analogy has been drawn frequently between such fragments and oceanic islands in the concept of the habitat island (MacArthur & Wilson, 1967; Simberloff & Abele, 1976; Diamond & May, 1976). There have been numerous studies which have demonstrated that the richness of species conforms to traditional species area and isolation relationships (reviewed by Spellerberg, 1991).

A very early example was the now-classic study by Moore (1962) on the Dorset heaths. He selected pairs of species - a pair of birds, of reptiles, of butterflies and of dragonflies. One member of each pair was a heathland specialist while the other occurred more generally. He showed that the heathland member of each pair tended to be absent from the isolated fragments of heathland, with the effect of isolation being greater than that of area. He considered that the absences were due to greater rates of extinction of species on the isolated heaths and a failure to colonize from other heaths. This resulted in an impoverishment of the faunas of the outlying heaths. He further predicted that over the following thirty years these trends would continue. Recent analyses (N.R. Webb, unpublished) have confirmed Moore's predictions: the faunas of the smaller, more isolated heaths have continued to decline, and the turnover of species has been greater on the smaller heaths than on the large.

Fragment area (as it has been defined above) works well as a unit for examining biodiversity for plants and some vertebrates (especially birds). However, when the relationships between heathland fragment area and isolation were examined for invertebrates on the Dorset heathlands (Webb & Hopkins 1984), no clear positive relationships between species richness and area were found. Hopkins and Webb (1984) were able to show that for a set of heathland spiders there were more species on large fragments of heathland than on small (but only a weak relationship), and that they occurred in greater abundance on the large fragments. There were no species of heathland spider present on small fragments that were not present on large ones. The species present on both large and small heathlands were shown to have greater dispersal abilities than those confined to large heathlands. The species with poor dispersal abilities were restricted to large heaths, where they persisted longer. This confirmed the view of Moore (1962) that the faunas of the more isolated fragments tended to be impoverished through a failure of species to recolonise. Hopkins & Webb (1984) speculated that following a local extinction on an area within a large heath, species may have been able to re-establish in an area more easily than on a small heath because the source of new colonists was from within the same fragment rather than from neighbouring heaths. Many of the fragments from which the colonists originate will be of a similar size to the fragment being colonised. This configuration of patches contrasts with the idea of a continent supplying colonists to off-shore islands. On large fragments, recolonisation of an area from which a species becomes extinct is likely to be from within the fragment.

Fragments of semi-natural vegetation often need to be managed to arrest succession. Where management is at a constant intensity over a long period, the structure and composition of a particular patch will remain the same. Because it is often easier, the management may be intermittent. An important feature of habitat management of this type is that it assumes there will be local extinction of species, followed by recolonisation from the surrounding populations or from a surviving section of a population. In effect, there has been an assumption of patch dynamics in the conservationists' concept of rotational habitat management, and an assumption of the long-term requirements of metapopulations in the concept of the integrated management of suites of reserves containing the same biotope.

The problem with the species-area approach

The studies of insects on heathland fragments have raised a number of interesting questions about how we should view such a patchwork. First, most studies of the relationships between the number of species and area, whether islands of habitat or real islands, have involved vertebrates, usually birds or higher plants. For all of these it is usually possible to make a census - to record every individual of each species present on the area. For insects this is impracticable, if not impossible, and the presence of species is ascertained by sampling. Because of this problem, Webb & Hopkins (1984) made comparisons of point diversity between heaths of different areas.

Secondly, although the fragments have been defined at what is effectively the human scale, this has been roughly suitable for analyses of species-area relationships in vertebrates, plants and certain other groups which in effect see the same configuration of patches as we do. Conservation science has interested itself considerably in this topic because it provides a means by which areas could be chosen to receive protection (May, 1975; Margules, Higgs & Rafe, 1982; Reed, 1983). In general, the greater the number of species within a given area, the more highly has the area been valued. There has been much discussion on the size, configuration, and whether several small areas conserve more species than a single large area (the SLOSS argument) (e.g. Gilpin & Diamond, 1980; Wilcox & Murphy, 1985; Quinn & Harrison, 1988; Spellerberg, 1991).

It is understandable that we should wish to relate the distribution and abundance of species to the pattern of fragments which exist. These are, after all, the practical units at which conservation operates: it is the scale at which humankind sees the fragments of semi-natural communities, and it is the scale at which they are owned and managed. This scale may be appropriate for other vertebrates, and indeed for some higher plants; however, it is not an appropriate scale for invertebrates, and for other species it may well be more appropriate to view the landscape at a relevant scale to the organisms under consideration. Likewise, while ownership may continue at the scale of fragments, conservation management will need to be on an appropriate scale within the boundaries of the fragments.

Thirdly, the essence of conservation is to attach a value to species, not all of which have the same claim to be conserved, and management or protective measures are directed according to species value. The approach of species-richness allied with island biogeography suggests that large areas should be protected because, thereby, the greatest number of species will be conserved and management easiest. This approach considers all species to have an equal value unless the index of biodiversity is weighted in some way (Vane-Wright *et al.*, 1991; Cousins, 1991). In fact, in many instances we are concerned to conserve a particular set of species, often those characteristic of a community, and we are concerned with what I have called representative diversity (Webb, 1989). That is, diversity is calculated by giving the highest weight to species characteristic of a particular community.

Habitats within the biotope

As an alternative to using species richness as a criterion, Webb & Thomas (1994) have adopted an approach based on consideration of the habitats of individual species. As a result of detailed studies on the Dorset heaths (Webb & Haskins, 1980; Chapman *et al.*, 1989; Webb, 1990), we are in a position to

i) define precisely the habitats of individual species,
ii) consider how well the heathland biotope provides these habitats, and
iii) consider how the heathland biotope could be modified to make better provision for habitats.

For this approach, it is important to recognise the very clear distinction between habitat and biotope. The fragments of heathland are the biotope, and I shall refer to them as heathland biotope fragments or a similar term. Within these biotope fragments are to be found the habitats of heathland species, and I shall refer to habitat patches in the narrow sense. This definition diverges from much recent literature on this subject, and I reserve the term "habitat" for the locality which is suitable for occupation by a particular species.

The general view of a patch of heathland gives no impression of the quantity of habitat available for an insect. On heathland, some general statements may be possible from an examination of the height or age of the heather bushes or a stand. This provides an approximate assessment of complexity (= plant structure). Likewise, an examination of the pattern of distribution of plant species, such as changes in the relative proportions of *Calluna vulgaris* and *Erica tetralix* (= an indication of the degree of wetness), provides an approximate measure of heterogeneity. What we are observing is the biotope and the conservationist may assume that the presence of heathland is alone sufficient to provide habitat to all species of interest: this may not be the case.

Thus, for the insects we have been studying we have defined the area which a population occupies as its habitat. For non-migratory, colonial species, this is a homogeneous area in a physical or abiotic sense and wholly occupied by the species. This area may remain in a condition suitable for occupation by the species for a variable length of time and may be linked, through dispersal, to adjacent similar areas. The occupation of the area by the species may not be continuous, extinctions occurring from time to time; re-establishment of the species on the area will be from adjacent areas, which may or may not be on the same biotope patch. The habitat patch often operates as a component of a metapopulation (Gilpin & Hanski, 1991). The availability temporally and spatially of habitat patches, their distribution, continuity and abundance, and occupancy are components of the habitat templet of Southwood (1977).

Examples from the Dorset heathlands

First, Webb & Thomas (1994) defined the habitat of selected heathland species using existing published information from autecological studies. This information was combined with that collected during surveys of heathland in Dorset in 1978 and 1987 (Webb & Haskins, 1980; Webb, 1990). By using the attributes in the Dorset heathlands survey database in this way, it was possible to produce maps which showed the distribution of the habitat patches for each species in relation to the distribution of the 141 fragments of heathland (biotope fragments). Thus, for each of the species we were able to examine the distribution of this unit of scale (the habitat patch) over the entire landscape, and to measure the number, size, distribution and persistence in time of the habitat patches, and to examine whether the number and size of the habitat patches within the heathland fragments (biotope patches) was related to the size of the heathland fragment.

The results (Webb & Thomas, 1994) showed that, despite the area of heathland in Dorset which is managed for conservation, the biotope has failed to provide habitat for either species with exacting habitat requirements (such as the ant *Myrmica sabuleti)* or for less demanding species (such as the Silver-studded Blue butterfly (*Plebejus argus*)). For the ant, the number of habitat patches became fewer and smaller over the nine-year period, and new patches of habitat were generated on average 9-10km away from the existing patches of habitat, thus making it very difficult for the ant to track the occurrence of its habitat. For the butterfly the situation was much better, and on average new habitat was created 80m from existing patches of habitat. Although this is not good, it is sufficient for the butterfly to track the occurrence of new habitat; nevertheless, only about one-third of the new habitat patches were near enough to existing patches for the butterfly to colonise them.

The next question is whether a better configuration of habitat patches is possible on these fragments of the heathland biotope. After all, this would be the aim of conservation management; however, the maximum amount of habitat might be restricted by topography, soils and hydrology. By using the definition of the habitat, it is possible to construct maps showing the potential habitat, were the heathland vegetation to be managed appropriately. From such maps, we have estimated that the currently-occupied habitat is only about 44% of the total that could be available if all the heaths were managed to provide the maximum possible area of habitat for the Silver-studded Blue.

Metapopulations

This approach is concerned with the availability of patches of habitat for heathland species within the fragments of the heathland biotope. The distribution of the habitat patches is constrained by the way in which the heathland fragments are distributed in the landscape. Again, we can use the Silver-studded Blue as an example. We assume that patches of habitat within 1km of existing colonies of the butterfly can be colonised over the time period for which they remain suitable (8-10 years) (Thomas & Harrison, 1992). Using the maps of the distribution of heathland fragments, we can join them together using a 1km rule to create what might be called matapopulations; thus one assumes that there is some interchange of individuals between all colonies which lie within 1km of each other. When this exercise was undertaken on the Dorset heaths over the nine-year period 1978-87, it was found that there were 19 metapopulations, 13 of which were very small. We can also make this calculation using the potential habitat, in which there are 13 metapopulations, nine of them small and isolated, seven of them consisting of a single component population, and the remaining four being very large with up to 37 constituent populations.

Large or small fragments?

Previously, it has been argued that only the largest fragments of heathland should be protected. This reasoning is based on the application of species area relationships, and the fact that heathland fragments are subject to very strong edge effects (Webb & Vermaat, 1990). The metapopulation approach alters our view of the patchwork of fragments. Strategically placed small heaths will provide links, if they are correctly managed. Likewise, there is a case for managing critical portions of the edges of large heaths to provide the habitats of heathland species and to maintain linkage with adjacent fragments.

In conclusion

This review of research on the dynamics of fragments of heathland in Dorset (southern England) has illustrated a number of important points which are of relevance to the way in which we view a landscape of this type.

Viewed from the landscape scale, the fragments of heathland appear to be stable; however, at the scale of the insect this is not the case, and each heathland fragment (biotope fragment) is seen to contain varying amounts of habitat for each species, and this habitat is present for varying periods of time. Hence, it is clear that we should define our fragments at a scale appropriate to the organism concerned and not use arbitrarily-defined fragments, such as those resulting from topographic mapping procedures. There is a clear need to distinguish between the biotope, which we might consider to be the habitat of the heathland community, and the habitat, which is a term that should be reserved for use in relation to a named species. The habitat is the place where a species is to be found, or perhaps more correctly a place which has all the features which would enable it to be occupied by the species, for not all of the available habitat for a species need be occupied. The biotope can be considered as the "habitat" of the community. Assemblages of species can generally be recognised, and the sum of their habitat requirements constitutes the biotope.

Traditional approaches, using theories of biogeography, have argued for the protection of the largest fragments of a biotope (Webb & Hopkins 1984) and those the least isolated (Moore 1962). However, when seen from the perspective of the individual species, there is a dynamic associated with the occurrence of habitat patches: when this is incorporated into the analysis, the importance of small fragments of the biotope is revealed and provides us with an alternative way of viewing a fragmented landscape and developing conservation policies and management.

References

Bullock, J.M. & Webb, N.R. (in press) Responses to severe fires in heathland mosaics. *Biological Conservation*.

Chapman S.B, Clarke R.T. & Webb, N.R. (1989) The survey and assessment of heathland in Dorset, England for conservation. *Biological Conservation* **47**: 137-152.

Cousins, S.H. (1991) Species diversity measurement: choosing the right index. *Trends in Ecology and Evolution* **6**: 190-192.

Diamond, J.M. & May, R.M. (1976) Island biogeography and the design of natural reserves. R.M. May (Ed.). *Theoretical Ecology, Principles and Applications*, Blackwell Scientific Publications, Oxford

Forman, R.T.T. & Godron, M. (1986) *Landscape Ecology*. Wiley, New York.

Gilpin, M.E. & Diamond, J.M. (1980) Subdivision of nature reserves and the maintenance of diversity. *Nature* **285**: 567-568.

Gilpin, M.E. & Hanski, I. (1991) *Metapopulation dynamics: empirical and theoretical investigations*, Academic Press, London.

Hopkins P.J. & Webb N.R. (1984) The composition of the beetle and spider faunas on fragmented heathlands. *Journal of Applied Ecology* **21**: 935-946.

MacArthur, R.H. & Wilson, E.O. (1967) *The Theory of Island Biogeography*. Princeton University Press, Princeton.

Margules, C.R., Higgs, A.J. & Rafe, R.W. (1982) Modern biogeographic theory: are there lessons for nature reserve design? *Biological Conservation* **24**: 115-128.

May, R.M. (1975) Patterns of species abundance and diversity. M. Cody & J.M. Diamond (Eds). *Ecology and evolution of communities*, Belknap Press, Harvard.

Moore N.W. (1962) The heaths of Dorset and their conservation. *Journal of Ecology* **50**: 369-391.

Quinn, J.F. & Harrison, S.P. (1988) Effects of habitat fragmentation and isolation on species richness: evidence from biogeographical patterns. *Oecologia* **75**: 132-140.

Reed, T.M. (1983) The role of species-area relationships in reserve choice: a British example. *Biological Conservation* **25**: 263-271.

Simberloff, D.S. & Abele, L.G. (1976) Island biogeography and conservation practice. *Science* **191**: 285-286.

Southwood, T.R.E. (1977) Habitat, the templet for ecological strategies. *Journal of Animal Ecology* **46**: 337-365.

Spellerberg, I.F. (1991) Biogeographical basis for conservation. I.F. Spellerberg; F.B. Goldsmith & M.G. Morris (Eds). *The Scientific Management of Temperate Communities for Conservation*. Blackwell Scientific Publications, Oxford.

Thomas C.D. & Harrison S. (1992) The spatial dynamics of a patchily distributed butterfly species. *Journal of Animal Ecology* **61**: 437-446.

Vane-Wright, R.I, Humphries, C.J. & Williams, P.H. (1991) What to protect? - systematic and the agony of choice. *Biological Conservation* **55**: 235-254.

Webb N.R. (1989) Studies on the invertebrate fauna of fragmented heathland in Dorset, UK, and the implications for conservation. *Biological Conservation* **47**: 153-165.

Webb N.R. (1990) Changes on the heathlands of Dorset, England, between 1978 and 1987. *Biological Conservation* **51**: 273-286.

Webb N.R. & Haskins L.E. (1980) An ecological survey of heathlands in the Poole Basin, Dorset, England in 1978. *Biological Conservation* 17, 281-296.

Webb N.R. & Hopkins P.J. (1984) Invertebrate diversity on fragmented *Calluna*-heathlands. *Journal of Applied Ecology* **21**: 921-933.

Webb, N.R. & Thomas, J.A. (1994) Conserving insect habitats in heathland biotopes: a question of scale. P.J. Edwards; R.M. May & N.R. Webb (Eds). *Large-scale Ecology and Conservation Biology*, Blackwell Scientific Publications, Oxford.

Webb, N.R. & Vermaat, A.H. (1990) Changes in vegetational diversity on remnant heathland fragments. *Biological Conservation* **53**: 253-264.

Wilcox, B.A. & Murphy, D.D. (1985) Conservation strategy: the effects of fragmentation on extinction. *American Naturalist* **125**: 879-887.

Narrow is the way

D.G. Dawson

London Ecology Unit, 125 Camden High Street, London NW1 7JR

Abstract

A review of the theoretical basis and empirical evidence for habitat corridors revealed few good studies. Corridors are supposed to lead animals and plants to places where they would not otherwise be found. The studies show that corridors sometimes can:- allow individual animals to survive by giving them access to sufficient habitat to meet their needs; serve the needs of some migratory animals in their seasonal movements; and perhaps maintain isolated populations of some animal and plant species by replenishment. However most species probably either fail to use corridors or can cross the gaps between patches of habitat adequately without their aid. It is doubtful whether good corridors can be found in many geographic regions to allow sensitive species to move in response to climate change. The idea that a network of corridors enables large-scale movement is attractive, but untested. It is recommended that existing corridors be kept and that corridors be provided where resources permit.

Introduction

Habitat corridors to lead animals and plants to places they would not otherwise be found are a popular concept with those designing or maintaining the landscape for nature conservation (Diamond, 1975; IUCN, 1980; Forman & Godron, 1986; Adams & Dove, 1989; Saunders & Hobbs, 1989; Verkaar, 1990; EEC, 1992). Despite this popularity, there has been little examination of the evidence for animals and plants actually depending upon corridors, and there is even less good evidence on just which species might use corridors of which kind.

Most of the support for a corridor function is purely theoretical, or from empirical findings with equivocal conclusions. Many papers are little more than anecdotal speculations, or refer to others who have advocated corridors, allowing enthusiastic endorsement of the "principle" (Harris & Scheck, 1991) as well as warnings that it may have little basis, or that resources may be employed better elsewhere (Simberloff & Cox, 1987). This literature review was commissioned by English Nature to explore the theoretical and empirical support there is for the function. The full review (Dawson 1994) is available from English Nature.

The theory

Much of the evidence for corridors comes from the large literature on "island biogeography", applying aspects of MacArthur & Wilson's (1967) mathematical model to situations where the habitat of animals and plants has been fragmented into patches isolated within an inhospitable matrix. Their examination of the role of "stepping stone" habitat patches in promoting immigration was the first mathematical model of the corridor function. More recently the theoretical focus has moved towards "metapopulation" (Levins, 1970; Wilcove *et al.*, 1986; Hanski, 1989; Taylor, 1991)and landscape ecology (Forman & Godron, 1981, 1986; Baudry 1984; Noss & Harris, 1986; Turner, 1989; Hansson & Angelstam, 1991; Opdam, 1991; Merriam, 1991).

Applications of these theories are dealing with situations where an animal or plant is supposed to:

- Be **specialised** for a habitat that was once widespread, but has been so depleted that it now occurs as isolated and small patches in an inhospitable matrix.

- **Disperse** so **poorly** that it is unable to replenish isolated populations to prevent their extinction, or to recolonise isolated patches where extinction has occurred.

Appropriate corridors are seen as reducing the isolation of patch populations by enhancing dispersal. In the extreme, a network of corridors and patches would act as a single, much larger patch (Merriam, 1991; Hobbs 1992), or a "minimum dynamic area" (Pickett & Thompson 1978). This is the "recolonisation" function of corridors (Dawson, 1994).

Developments of the basic theory have covered such aspects as the attributes of species that might require corridors and the desirable width and habitat composition of corridors. Species that might require corridors are those that are susceptible to extinction. In terms of MacArthur's (1972) "r-K" continuum, or Southwood's (1977) "habitat templet" these should be adapted to extensive, stable (often "climax") habitats. In comparison with related species, these susceptible species tend to be large, have a low rate of reproduction, low population density and poor dispersal ability. Stepping stone theory (Gilpin 1980), the well-studied species-area effect (Hart & Horwitz, 1991), the correlation between species and habitat diversities (Western & Ssemakula, 1981; Game & Peterken, 1984; Simberloff & Abele, 1984) and studies of adverse habitat conditions at the edge of habitat patches (Forman & Godron, 1981; Start, 1991; Merriam & Wegner, 1992), suggest that corridors are better if they are as continuous and wide as possible.

Much of this theory, however, applies strictly to the **recolonisation** function, whereas there are four other situations where corridors may prove advantageous (Dawson, 1994) and for the first two of these, at least, a colonisation-extinction model is inappropriate:-

- **Size threshold.** Corridors may allow individual animals to move between two or more patches to find enough habitat for day-to day survival where one patch would be too small to support them (Sullivan & Schaffer, 1975; Hobbs, 1992).

- **Migration.** Animals may be enabled to move to and from the different habitats they require each season, and so find their requirements for survival (Harris & Scheck, 1991).

- **Climate change.** Species of animal or plant may need to follow their habitat as it changes its distribution in response to changing climate (Hill et al., 1994).

- **Incidence.** Corridors may simply enable species or individuals which are not at any risk of extinction or death to be seen more widely, or more often, than they otherwise would be, by giving them more ready access to their habitat.

The empirical support in biogeography

Much of the empirical support for corridors comes from studies of animals, particularly birds. The studies suggest that populations are specialised, isolated, disperse poorly and have been lost from remote or small fragments (Dawson, 1994); the need for corridors is deduced without studying actual corridors (Robbins *et al.*, 1987; Soule *et al.*, 1988; Turner, 1989). This general literature on island biogeography or habitat fragmentation is not without its problems (Gilbert, 1980; Haila, 1983; Boecklen & Gotelli, 1984; Boecklen & Simberloff, 1986; Burgman *et al.*, 1988; Opdam, 1990; Harrison, 1991), but the better studies do suggest that some species are at risk from habitat fragmentation and that corridors might ease the problem (Dawson, 1994). The studies also suggest that wide corridors will be better than narrow ones.

An examination of the dispersal rates of species shows that many do not require corridors, because they are physically and behaviourally able to cross the inhospitable matrix without assistance (den Boor, 1990; Bennett, 1991a; Dawson 1994). Others (such as clonal woodland herbs) disperse, even through favourable habitat, so slowly that colonisation is impossible in any realistic time (Peterken, 1993). These two groups of species are the majority, and can benefit little, if at all, from corridors. This view is based upon a review of the movement abilities of species (Dawson, 1994) and also on the difficulty of demonstrating a convincing effect of the degree of isolation on species number in most studies (Connor & McCoy, 1979; Hart & Horwitz, 1991).

However, there ought to be an intermediate, if small and specialised, group of species which are able to move sufficiently well through corridors of suitable habitat, but are unable to cross the matrix successfully (Dawson, 1994).

Studies of barriers (such as roads) have documented that some animal species may rarely cross, or incur high mortality in doing so (Merriam *et al.*, 1989; Mader *et al.*, 1990; Bennett, 1991a & b). These findings suggest that barriers may slow recolonisation, but they are not enough to show that they prevent it, as even a rare successful crossing may be sufficient to replenish an isolated patch. However, for individual animals to achieve their habitat size threshold, or for migrants to complete their seasonal movements, every animal must have a good probability of survival, and so the findings do show that corridors could be effective in these two respects.

Studies of actual corridors

Unfortunately there have been very few good studies of actual corridors. Just as in the studies of island biogeography and habitat fragmentation (see above), most studies lack the full requirements of hypothesis testing: comparison with adequate "control" situations, replication, and rejection of a null hypothesis in a statistical test. Indeed some did not even present data in support of their conclusions (Dawson, 1994).

A good number of studies of animal movement have shown that it occurs much more often in corridors than the matrix (Saunders, 1990; Hobbs, 1992; Dawson 1994). These studies lacked sufficient unconnected "control" areas and so, while they suggest, they do not prove that recolonisation would be impeded seriouslywithout corridors. As with the evidence on barriers (above), these are adequate, however, to show that corridors can provide a size threshold or migration route, especially for movements of terrestrial animals, such as amphibia or migratory mammals of the seasonal tropics and the arctic.

Only one study was found that met the requirements of hypothesis testing: that by Pollard *et al.* (1974) on the distribution of dog's mercury, *Mercurialis perennis,* in six planted hedgerows extending from a Northamptonshire wood. This study is interesting, also, because the rate of movement was insufficient to serve the human desire to see results within a lifetime!

Is the evidence enough?

The review has shown that the preconditions for corridors to act as useful conduits can occur, but that the evidence for recolonisation comes from theory, studies of movements, statistically inadequate work, and only one good empirical study of corridors themselves. This is slim support for an important concept in practical conservation and landscape ecology; there have been very few adequate studies and the review provides suggestions for further work to remedy this (Dawson, 1994).

Corridors for species to move in response to climate change must extend even further than those for recolonisation and it seems less likely that suitable corridors for the sensitive species can be found, even though the species have a long time to complete the passage.

In contrast, the evidence from movement along corridors, and not across barriers, confirms the utility of appropriate corridors to provide some animal species with their area threshold for survival, or with an essential migration route.

The precautionary principle (O'Riordan, 1992) means that we cannot await the completion of the studies that would be required to confirm these findings across the wide range of circumstances where they may apply. Hobbs (1992) reminded us of the ease with which corridors can be retained in contrast to the prospect of replacing them if lost. The practical conclusion, then, is that habitat corridors that connect other nature conservation sites and lead into the inhospitable surrounds, should be kept, and that they should be improved, or created where resources permit. This is partly because they would be effective for some animals, and probably plants, and also because we cannot await proof of which species these are (Saunders *et al.*, 1991; Harris & Scheck, 1991). At the least these corridors should serve to enhance the incidence of species which are not threatened with extinction and allow them to be appreciated in a wider range of places.

Postscript - a network of corridors in the landscape

Landscape ecologists have theorised that a network of corridors and patches may enable movement over large distances through the landscape (Turner, 1989; Opdam, 1991; Merriam & Wegner, 1992). This is an attractive idea, but it depends upon either a demonstration that all the individual corridor links are effective (and that is the subject of the rest of the review; Dawson, 1994), or developing measures of connectivity to use at the landscape scale which discriminate between the conduit function of corridors and their contribution of habitat regardless of species movement. The review found no studies meeting the latter requirement (Dawson, 1994).

Strait is the gate, and narrow is the way, which leadeth unto life, and few there be that find it.

Matthew vii. 14.

Acknowledgements

I am grateful to Keith Kirby and George Barker of English Nature for assistance at every stage of this review, and to very many others who are acknowledged in the larger paper (Dawson, 1994).

References

Adams, L.W. & Dove, L.E. (1989) *Wildlife reserves and corridors in the urban environment, a guide to ecological landscape planning and resource conservation.* National Institute for Urban Wildlife, Columba, Maryland.

Baudry, J. (1984) Effects of landscape structure on biological communities: the case of hedgerow network landscapes. *Methodology in Landscape, Ecological Research and Planning* **1**:55-65.

Bennett, A.F. (1991a) What types of organism will use corridors? D.A. Saunders & R.J. Hobbs (Eds). *Nature Conservation 2: the role of corridors.* Surrey Beatty, Chipping Norton, NSW 407-408.

Bennett, A.F. (1991b) Roads, roadsides and wildlife conservation: a review. D.A. Saunders & R.J. Hobbs (Eds). *Nature Conservation 2: the role of corridors.* Surrey Beatty, Chipping Norton, NSW 99-108.

Boecklen, W.J. & Gotelli, N.J. (1984) Island biogeographic theory and conservation practice: species-area or specious-area relationships? *Biological Conservation* **29**:63-80.

Boecklen, W.J. & Simberloff, D. (1986) Area-based extinction models in conservation. D.K. Elliott (Ed). *Dynamics of extinction,* Wiley, New York 247-276.

Burgman, M.A.; Akcakaya, H.R. & Loew, S.S. (1988) The use of extinction models for species conservation. *Biological Conservation* **43**:9-25.

Connor, E.F. & McCoy, E.D. (1979) The statistics and biology of the species-area relationship. *American Naturalist* **113**:791-833.

Dawson, D.G. (1994) *Are habitat corridors conduits for animals and plants in a fragmented landscape? A review of the scientific evidence.* English Nature, Research Report 94, Peterborough.

den Boer, P.J. (1990) The survival value of dispersal in terrestrial arthropods. *Biological Conservation* **54**:175-192.

Diamond, J.M. (1975) The island dilemma: lessons of modern biogeographic studies for the design of natural preserves. *Biological Conservation* 7:129-146.

European Economic Community. (1992) *On the conservation of natural habitats and wild fauna and flora.* EEC Council Directive 92/43.

Forman, R.T.T. & Godron, M. (1981) Patches and structural components for a landscape ecology. *Bioscience* **31**:733-740.

Forman, R.T.T. & Godron, M. (1986) *Landscape ecology*. John Wiley & Sons, New York.

Game, M. & Peterken, G.F. (1984) Nature reserve selection strategies in the woodlands of central Lincolnshire, England. *Biological Conservation* **29**:157-81.

Gilbert, F.S. (1980) The equilibrium theory of island biogeography: fact or fiction. *Journal of Biogeography* 7:209-235.

Gilpin, M.E. (1980) The role of stepping-stone islands. *Theoretical Population Biology* **17**:247-253.

Haila, Y. (1983) Land birds on northern islands: a sampling metaphor for insular colonization. *Oikos* **41**:334-51.

Hanski, I. (1989) Metapopulation dynamics: does it help to have more of the same? *Trends in Ecology & Evolution* 4:113-114.

Hansson, L. & Anglestam, P. (1991) Landscape ecology as a theoretical basis for nature conservation. *Landscape Ecology* **5**:191-201.

Harris, L.D. & Scheck, J. (1991) From implications to applications: the dispersal corridor principle applied to the conservation of biological diversity. D.A. Saunders & R.J. Hobbs (Eds). *Nature Conservation 2: the Role of Corridors,* Surrey Beatty, Chipping Norton, NSW 189-220.

Harrison, S. (1991) Local extinction in a metapopulation context: an empirical evaluation. *Biological Journal of the Linnaean Society* **42**:73-88.

Hart, D.D. & Horwitz, R.J. (1991) Habitat diversity and the species-area relationship: alternative models and tests. S.S. Bell, E.D. McCoy & H.R. Mushinsky (Eds). *Habitat Structure. The Physical Arrangement of Objects in Space.* Chapman & Hall, London 47-68.

Hill, M.O.; Carey, P.D.; Eversham, B.C.; Arnold, H.R.; Preston, C.D.; Telfer, M.G.; Brown, N.J., Veitch, N. & Welch, R.C, (1994) *The role of corridors, stepping stones and islands for species conservation in a changing climate.* English Nature Research Report 75, Peterborough.

Hobbs, R.J. (1992) The role of corridors in conservation: solution or bandwagon? *Trends in Ecology and Evolution* **7**:389-392.

Sutton, R. & Vargas, T. (1987) Comparison of neotropical winter bird populations in isolated patches versus extensive forest. *Acta Oecologia, Oecologia Generalis* **8**:285-292.

Saunders, D.A. (1990) Problems of survival in an intensively cultivated landscape: the case of Carnaby's cockatoo, *Calyptorhynchs funereus latirostris. Biological Conservation* **54**:277-290.

Saunders, D. & Hobbs, R. (1989) Corridors for conservation. *New Scientist*, 28 Jan, 63-8.

Saunders, D.A.; Hobbs, R.J. & Margules, C.R. (1991) Biological consequences of ecosystem fragmentation: a review. *Conservation Biology* **5**:18-32.

Simberloff, D.S. & Abele, L.G. (1984) Conservation and obfuscation: subdivision of reserves. *Oikos* **42**:399-401.

Simberloff, D.S. & Cox, J. (1987) Consequences and costs of conservation corridors. *Conservation Biology* **1**:63-71.

Soule, M.E.; Bolger, D.T.; Alberts, A.C.; Wright, J.; Sorice, M. & Hill, S. (1988) Reconstructed dynamics of rapid extinctions of chaparral-requiring birds in urban habitat islands. *Conservation Biology* **2**:75-92.

Southwood, T.R.E. (1977) Habitat, the templet for ecological strategies? *Journal of Animal Ecology* **46**:337-365.

Start, A.N. (1991) How can edge effects be minimised? D.A. Saunders & R.J. Hobbs (Eds). *Nature Conservation 2: the Role of Corridors*, Surrey Beatty, Chipping Norton, NSW 417-418.

Sullivan, A.L. & Shaffer, M.L. (1975) Biogeography of the megazoo. *Science* **189**:13-17.

Taylor, A.D. (1991) Studying metapopulation effects in predator-prey systems. Biological Journal of the Linnean Society**42**:305-323.

Turner, M.G. (1989) Landscape ecology: the effect of pattern on process. *Annual Review of Ecology and Systematics* **20**:171-197.

Verkaar, H.J. (1990) Corridors as a tool for plant species conservation? R.G.H. Bunce & D.C. Howard (Eds). *Species dispersal in agricultural habitats,* Belhaven, London 82-97.

Western, D. & Ssemakula, J. (1981) The future of the savannah ecosystems: ecological islands or faunal enclaves? *African Journal of Ecology* **19**:7-19.

Wilcove, D.S., McLellan, C.H. & Dobson, A.P. (1986) Habitat fragmentation in the temperate zone. M.E. Soule (Ed). *Conservation Biology. The Science of Scarcity and Diversity.* Sinauer, Sunderland, Mass 237-256.

IUCN (1980) *World Conservation Strategy*. International Union for Conservation of Nature and Natural Resources, Gland, Switzerland.

Levins, R. (1970) Extinction. M. Gerstenhaber (Ed). *Some Mathematical Problems in Biology*, American Mathematical Society, Providence, Rhode Island 75-107.

MacArthur, R.A. (1972) *Geographical Ecology*. Harper & Row, New York.

MacArthur, R.A. & Wilson, E. (1967) *The theory of island biogeography*. Princeton University Press, Princeton, NJ.

Mader, H.-J.; Schell, C. & Kornacker, P. (1990) Linear barriers to arthropod movements in the landscape. *Biological Conservation* **54**:209-222.

Merriam, G. (1991) Corridors and connectivity: animal populations in heterogeneous environments. D.A. Saunders & R.J Hobbs, *Nature Conservation 2: The Role of Corridors*, Surrey Beatty, Chipping Norton, NSW 133-142.

Merriam, G.; Kozakiewicz, M.; Tsuchiya, E. & Hawley, K. (1989) Barriers as boundaries for metapopulations and demes of *Peromyscus leucopus* in farm landscapes. *Landscape Ecology* **2**:222-235.

Merriam, G. & Wegner, J. (1992) Local extinctions, habitat fragmentation, and ecotones. A.J. Hansen & F. di Castri (Eds). *Landscape boundaries. Consequences for biotic diversity and ecological flows*, Springer-Verlag, London 150-169.

Noss, R.F. & Harris, L. (1986) Nodes, networks and MUMs: preserving diversity at all scales. *Environmental Management* **10**:299-309.

Opdam, P. (1990) Dispersal in fragmented populations: the key to survival. R.G.H. Bunce & D.C. Howard (Eds). *Species Dispersal in Agricultural Habitats*, Belhaven, London 3-17.

Opdam, P. (1991) Metapopulation theory and habitat fragmentation: a review of Holarctic breeding bird studies. *Landscape Ecology* **5**:93-106.

O'Riordan, T. (1992) A new science for a new age. *New Scientist* 25 Jan:14.

Peterken, G.F. (1993) Landscape-scale issues in the conservation of ancient woodlands. R. Haines-Young (Ed). *Landscape ecology in Britain*, IALE(UK)/Department of Geography, University of Nottingham, Nottingham 55-65.

Pickett, S.T.A. & Thompson, J.N. (1978) Patch dynamics and the design of nature reserves. *Biological Conservation* **13**:27-37.

Pollard, E.; Hooper, M.D. & Moore, N.W. (1974) *Hedges*. Collins, London.

Robbins, C.S.; Dowell, B.A.; Dawson, D.K.; Colon, J.; Espinosa, F.; Rodrigrez, I.;

Extensive farmland in Europe: examples of sustainable landscapes of high nature conservation value

D.I. McCracken[1], R.A. Stillman[2], E.M. Bignal[3], D. Baldock[4] and G. Beaufoy[4]

[1] Joint Nature Conservation Committee, 25 High Calside, Paisley PA2 6BY, UK
[2] Joint Nature Conservation Committee, Monkstone House, City Road, Peterborough PE1 1JY, UK (Present address: Institute for Terrestrial Ecology, Furzebrook Research Station, Wareham, Dorset, BH20 5AS, UK)
[3] Joint Nature Conservation Committee, Kindrochaid, Bruichladdich, Isle of Islay, Argyll PA44 7PT, UK
[4] Institute for European Environmental Policy, 158 Buckingham Palace Road, London SW1W 9TR, UK

Abstract

Extensive farming systems are responsible for the creation and continuing maintenance of the landscapes and habitats currently required by many species of nature conservation importance in Europe. However, throughout Europe these systems are disappearing or breaking down. Measures need to be put in place urgently to help preserve the landscape and wildlife value of these systems while at the same time allowing the associated human communities to adapt the systems to modern needs. Complex relationships need to be understood before the social, environmental and economic consequences of any changes in policy can be assessed fully.

Introduction

The effects of the last five decades of agricultural intensification in reducing semi-natural habitats and the distribution of wildlife in Europe have been frequently documented. Consequently, it would be easy to believe that all present-day farming practices have adverse effects on wildlife and to condemn them accordingly. However, what is still not widely appreciated is that throughout Europe large tracts of land under less-intensive agricultural management (extensive farming systems) are of great nature conservation importance for the wildlife assemblages they contain. Indeed, many of the wildest and most remote parts of Europe are in fact extensive farmland, and agricultural activities have played a major role in the development and maintenance of the landscapes and semi-natural habitats on which we now place such high value.

It is now accepted that agricultural policies must take more account of the environment, and a significant new aid programme (Regulation 2078/92) has been included in the European Union (EU) Common Agricultural Policy. However, there is still little detailed information available on the character, distribution and

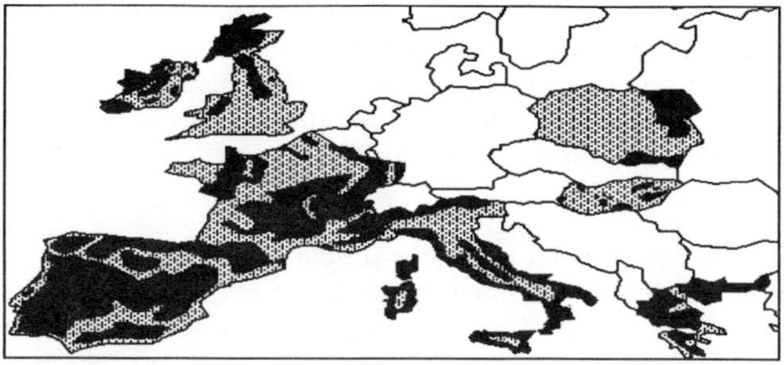

Figure 1 Principal areas (black) within which extensive farming systems can potentially be found in each of the nine European countries (shaded) considered by Baldock *et al.* (in press).

nature conservation importance of European extensive farming systems, nor of the changes which they are currently undergoing. The main purpose of the work reported here is to contribute towards a greater understanding of these systems, and to thereby ensure that policies for their maintenance and development are formulated in such a way that their nature conservation value is protected and enhanced.

Extensive farming systems in Europe

A recently completed pilot study has provided information at a broad-scale on the extent, distribution and characteristics of existing extensive farming systems in nine European countries (Baldock *et al.*, in press).

Extent and location

Figure 1 shows the principal areas within which extensive systems are potentially still being practised. Most of the farmland which is still extensively managed occurs in areas where severe physical constraints limit intensification, particularly in upland and mountainous areas, arid zones and the relatively small area of wet soils which have not yet been drained. In some regions the constraints have been less physical than socio-economic, for example the land may be inaccessible, remote from the market or in such fragmented ownership that intensification is impractical. The greatest concentration of extensive farmland is found in the Mediterranean region, particularly the Iberian Peninsula.

Character and conservation value

The great majority of extensive farming systems in Europe involve the exploitation of grass or semi-natural vegetation as pasture for domestic livestock. Such pastoral systems range from semi-wild and largely unmanaged cattle and horses in remote mountain regions of Spain and Italy to highly commercialised dairy cattle systems based on closely managed meadows in the French Jura. Extensive sheep systems are the most widespread and can be found in almost every country. The sources of forage include grass-dominated pastures and meadows, coarser forms of semi-natural vegetation, and also woodland in some areas.

Although much less widespread than extensive grazing systems, a variety of extensive cropping or mixed farming systems can also still be found. These include low input arable systems (predominately in the drier Mediterranean climates), mixed cropping and crop/livestock systems such as cultra promiscua in Italy or crofting in northern Scotland, and extensive orchards and olive groves in Spain, France and Italy.

Baldock *et al.* (in press) have developed a relatively simple typology to present the extensive systems according to broadly defined categories of agricultural land use. They have thus identified eight mainly livestock systems (L1 to L8), four mainly arable systems (A1 to A4), two permanent crop systems (P1 & P2) and two mixed systems (M1 & M2). Details of the character and conservation value of two of the livestock systems are given below.

Livestock raising in wooded pastures (L5)

This involves extensive grazing of livestock on pastures under trees at a density varying between open woodland and scattered individual trees. Systems of this type are practised in small areas of north-east Hungary, south-east Poland and central Italy. However, the largest area occurs in the south-west of the Iberian Peninsula, where livestock graze pastures mainly under a mixture of cork oak (*Quercus suber*) and holm oak (*Q. rotundifolia*) trees. Small areas of cereals may also be cultivated (in rotation with fallow and pasture) under the trees to provide supplementary forage for the livestock. These wooded pastures are called dehesas in Spain and montados in Portugal. The oak trees provide shelter for the livestock and crops and help to hold water in the soil, but are also essential productive components of the dehesa and montado systems. For example, the acorns form an essential part of the diet of the pigs which graze these areas, cut branches provide extra fodder for the livestock in the winter months and cork is harvested from the cork oak trees.

Traditional management of dehesas and montados maintains a characteristic landscape containing elements of woodland, species-rich grasslands and scrub. This mixture of habitats contains a remarkable abundance and variety of wildlife, e.g. mammals such as the wolf (*Canis lupus*) and Iberian lynx (*Lynx pardina*), reptiles such as the laddersnake (*Elaphe scalaris*) and Montpellier snake (*Malapon monspessulanus*), and especially a large number of bird species. Dehesas and montados are essential to the survival of threatened species such as the imperial eagle (*Aquila heliaca*) and black stork (*Ciconia nigra*), and provide important wintering areas for millions of northern European migrants such as the common crane (*Grus grus*), black-cap (*Sylvia atricapilla*) and chiffchaff (*Phylloscopus collybita*).

Transhumance and other seasonal movements of livestock between grazing areas (L8)

Transhumance systems can take many different forms, from short movements of livestock between different types of grazing land within a region according to season, to very long distance movements between lowland (winter) and upland (summer) regions. Today long distance transhumance usually only involves sheep and goats, while shorter distances can also involve cattle and sometimes horses. These systems are now mainly practised in the southern European countries such as Spain (where livestock go from the warmer and drier central and lowland areas to the mountainous regions during the summer), Italy (where livestock winter on the lowland areas and spend the summer in the Alps or the Apennines) and Greece (where the main movements are between the eastern and western lowlands and the Pindos mountains in the centre of the country).

Transhumance is essential to the survival of many habitats of high landscape and wildlife value (e.g. species-rich mountain pastures and meadows), as grazing for a relatively short period each year at relatively low stocking densities is necessary to maintain these habitats and their floristic diversity. In addition, grazing by livestock and the presence of carcasses maintains the open areas with carrion that are essential for foraging by scavengers (such as lammergeier, *Gypaetus barbatus*, griffon vulture, *Gyps fulvus*, and red kite, *Milvus milvus*) and predators (such as golden eagle, *Aquila chrysaetos*, Bonelli's eagle, *Hieraaetus pennatus*, and wolves).

Vulnerability

As farming systems are dynamic and in a constant state of evolution, extensive farming systems should not be thought of as purely 'traditional' and unchanging. Indeed, many of the systems identified by Baldock *et al.* (in press) exhibited signs of development, and the combination of the different management techniques in use was found to be shifting in many regions. However, economic and social pressures are currently increasing the speed of change within many of these systems, and thereby putting their future, and that of their associated wildlife communities, under threat.

The principal developments threatening the survival of extensive systems are intensification, abandonment, and afforestation. For example, the area of dehesas and montados has decreased greatly in the last few decades mainly as a result of the establishment of large irrigation projects (allowing more intensive and mechanised farming practices) and large areas of plantation forestry (for timber extraction). Many of those dehesas and montados that remain have deteriorated due to the abandonment of traditional management practices or the lack of tree regeneration (often as a result of increased stocking densities). In addition, the abandonment of transhumance in many mountain areas has resulted in the deterioration of alpine and sub-alpine habitats through scrub encroachment. The livestock that used to be taken to the mountains each summer now remain in the lowlands (or lower down the mountains) throughout the year. This has also led to the introduction of different breeds and the over-grazing of some lowland habitats such as grassland steppes.

Detailed research in Britain

The majority of extensive farmland in Britain is now confined to the upland areas in the north and west of the country, and is predominately based on the extensive grazing of sheep and cattle and the production of small amounts of cereals and hay (or silage) for use as winter feed (Bignal & McCracken, 1993). Two recent surveys of the vegetation and breeding birds of extensive farmland in north-east England (mainly under systems L1 or L2) and western Scotland (mainly under systems L1 or M2), have provided some more specific details of the nature conservation value of such farmland. This information has been used to investigate both the extent of habitat fragmentation in these areas, and the influence of habitat on the distribution of bird species. Such information aids the understanding of the complex interactions involved, which is essential before appropriate policy measures can be formulated.

Habitat fragmentation

During the surveys, the vegetation was recorded as patches of homogeneous 'cover types' and by the relative abundance of individual species occuring in each patch. Cover types are expressed here in terms of Nature Conservancy Council Phase 1 habitats. Four of these (calcareous grassland, dry and wet dwarf shrub heaths, and bog) are equivalent to habitats or groups of habitats listed on Annex 1 of EU Directive 92/43 on the conservation of natural habitats of wild fauna and flora, and are therefore important in an international context.

The distribution of habitats within a sample of 1 x 1 km squares (74 in England and 61 in Scotland) was recorded by noting the dominant cover type in each cell of a 50 x 50 grid superimposed on each square. The relative cover of each habitat was then recorded from the number of occupied cells, and an index of fragmentation calculated from the proportion of occupied cells which were adjoining different habitats. The index could range from 0 (no fragmentation) when a single habitat occupied an entire 1 km square), to 1 (maximum fragmentation) when a habitat was divided into a number of patches each occupying a single cell and therefore completely bounded by cells of a different habitat.

The fragmentation of those major habitats which occured in both survey areas, was investigated by considering the relative cover of each habitat (throughout all the 1 km squares sampled within each area) together with the mean fragmentation index of each habitat type in occupied 1 km squares (Figure 2). Of the Directive 92/43 habitats recorded during the surveys, only wet heath and bog in Scotland were relatively abundant, with dry heath and calcareous grassland occurring relatively infrequently in both areas. Additionally wet heath in England, dry heath in Scotland, and calcareous grassland in both areas were relatively fragmented in comparison to the other habitat types recorded. Bog was unfragmented in both areas, as was dry heath in England and wet heath in Scotland.

The remaining habitat types recorded in the surveys similarly showed a range of abundance and fragmentation. In both areas flush and bracken were relatively infrequent and fragmented, and neutral grassland was abundant and unfragmented. Acid grassland was relatively more abundant and less fragmented in England than in Scotland.

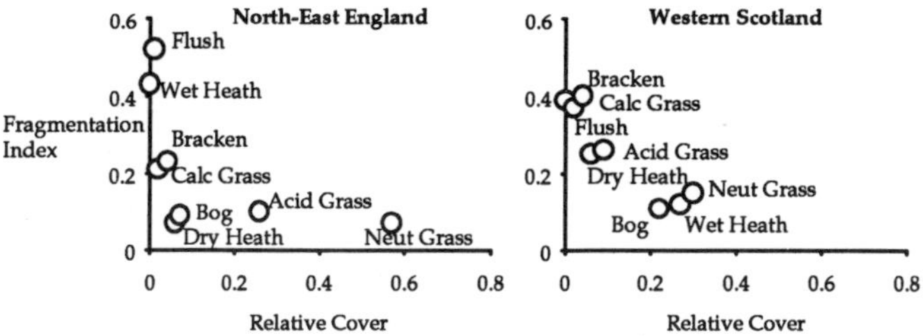

Figure 2 Habitat fragmentation in north-east England and western Scotland.

Habitat selection by breeding bird species

The analyses were confined to Red Data or Candidate Red Data bird species (as given in Batten *et al.*, 1990). In total 18 species of such birds were recorded breeding in north-east England and 22 species in western Scotland. Four waders (curlew, *Numenius arquata*, lapwing, *Vanellus vanellus*, redshank, *Tringa totanus*, and snipe, *Gallinago gallinago*), and two passerines (whinchat, *Saxicola rubetra*, and wheatear, *Oenanthe oenanthe*) were relatively abundant in both areas. These six species were therefore used to investigate habitat association using an index relating the proportion of records in a habitat to the frequency of occurrence of that habitat within each area. The bird species showed marked differences in their overall habitat associations (Table 1).

The bird species showed fewer preferences for the Directive 92/43 habitats than for the remaining habitats. For the two areas combined, nine positive associations were found between birds and the four Directive 92/43 habitats, while 21 were found for the remaining four habitats. Wet heath was not preferably selected by any of the six species studied, while dry heath was selected by curlew in both areas and by wheatear in Scotland. Bog was selected by curlew, lapwing, snipe and whinchat in Scotland, but by none of these species in England. Calcareous grassland was selected by wheatear in both areas.

Of the remaining habitats, bracken was selected by both wheatear and whinchat in both areas, and avoided by all of the wader species. Flush was selected by redshank and snipe in both areas, and by lapwing and wheatear in Scotland. Acid grassland was selected by wheatear and all wader species except lapwing in England, and by wheatear and whinchat in Scotland. Neutral grassland was selected by lapwing and redshank in both areas, and by wheatear in Scotland.

Conclusions

The report by Baldock *et al.* (in press) indicates that throughout Europe extensive farming systems are disappearing or breaking down. In some areas land is being abandoned, in others it is subject to agricultural intensification or entirely new uses such as commercial afforestation. Yet these systems are of particular environmental value, not only because they are less polluting and demanding in their use of

Table 1 Habitat association by bird species in extensive farmland. Values greater than 0.3 are marked in bold type and indicate positive selection for a habitat.

England				Habitat				
	dry heath	wet heath	bog	acid grass	neut grass	calc grass	flush	bracken
curlew	**0.38**	0.00	0.25	**0.36**	0.28	0.14	0.11	0.21
lapwing	0.15	0.00	0.10	0.24	**0.37**	0.16	0.00	0.00
redshank	0.25	0.00	0.00	**0.34**	**0.32**	0.00	**0.53**	0.00
snipe	0.07	0.00	0.16	**0.40**	0.29	0.00	**0.31**	0.00
whinchat	0.00	0.00	0.00	0.12	0.11	0.00	0.00	**1.39**
wheatear	0.16	0.00	0.09	**0.33**	0.25	**0.97**	0.00	**0.49**

Scotland				Habitat				
	dry heath	wet heath	bog	acid grass	neut grass	calc grass	flush	bracken
curlew	**0.39**	0.27	**0.38**	0.28	0.29	0.00	0.00	0.00
lapwing	0.14	0.13	**0.35**	0.00	**0.41**	0.00	**0.85**	0.00
redshank	0.00	0.00	0.26	0.00	**0.52**	0.00	**0.76**	0.00
snipe	0.27	0.20	**0.49**	0.00	0.26	0.00	**0.66**	0.00
whinchat	0.00	0.10	**0.48**	**0.50**	0.20	0.00	0.00	**0.67**
wheatear	**0.43**	0.13	0.21	**0.42**	**0.38**	**0.38**	**0.51**	**0.41**

resources than intensive agriculture, but also because of their role in conserving landscapes, habitats and wildlife communities of recognised European importance.

The survey work conducted in two areas of extensive farmland in Britain has indicated that some of the habitat types recorded (such as flush, bracken and calcareous grassland) tend to be fragmented in both areas. While this fragmentation may be a natural state for some habitats (such as flushes), it may be an indication of degradation in others (such as wet and dry heath in north-east England). Initial analyses of the data have shown broad associations between bird species and habitats which are generally consistent with those recorded in other complementary studies (e.g. Brown & Stillman, 1993).

The effects of fragmentation on breeding bird habitat selection are more difficult to describe. For instance, preliminary analyses suggest that higher lapwing densities occured in unfragmented neutral grasslands, whereas wheatear densities showed no clear relationship with fragmentation. Clearly the influence of habitat fragmentation will differ depending on species ecology, and therefore value judgements about the implications of habitat fragmentation must relate to clearly defined ecological or landscape objectives. The analyses presented above indicate that management directed solely at Directive 92/43 habitats would not necessarily be favourable for any of the important bird species considered here.

No analyses have been possible for the two examples of extensive systems given above (L5 and L8), but it is suspected that fragmentation at the landscape level is increasing as a result of intensification of farming practices in southern European arid landscapes, and is decreasing as a result of abandonment of farming practices in mountain landscapes. Therefore nature conservation value is declining through an increase in fragmentation in some areas and a decrease in fragmentation in others.

Future changes in the European landscape will be strongly influenced by agricultural policies, and this could radically affect both the direction and rate of fragmentation, e.g. through incentives to abandon farmland. There is therefore a need to influence such policies to further nature conservation in the European countryside. However, it must always be remembered that ecological and conservation issues cannot be divorced from social, cultural and economic ones. Complex relationships therefore need to be understood and clear objectives need to be set before the social, environmental and economic consequences of any changes in policy can be assessed fully.

Acknowledgements

The study into the extent, distribution and characteristics of extensive farming systems in Europe was conducted by IEEP and funded by WWF International and the JNCC. Many people provided valuable assistance and encouragement during this study, in particular Jacques Baudry, Nico Beopoulos, Isabel Bermejo, Julian Clark, Heather Corrie, Isabelle Dubien, Jacques Gilbert, Anna Grontkowska, Bogdan Klepacki, Emmanuel Lierderman, Leonidas Louloundis, Ferenc Markus, Marta Paluch, Kate Partridge, Francesco Petretti, Mike Pienkowski, Xavier Poux, Zeno Tederko, Czeslaw Siekerski, Ludwig Wicki and Tanya Yudelman. The detailed research in Britain was initiated by the NCC (and subsequently continued by the JNCC), and thanks are due to Dave Curtis, Allan Drewitt, Chris Moos, Gy Ovenden, Mandy Wilson and all the field workers for their involvement with this project.

References

Baldock, D.; Beaufoy, G.; Bignal, E. & Clark, J. (in press) *Extensive Farming Systems in Europe*. Institute for European Environmental Policy, London.

Batten, L.A.; Bibby, C.J.; Clement, P.; Elliot, G.D. & Porter, R.F. (1990) *Red Data Birds in Britain*. Poyser, London.

Bignal, E.M. & McCracken, D.I. (1993) Nature conservation and pastoral farming in the British uplands. *British Wildlife* **4:** 367-376.

Brown, A.F. & Stillman, R.A. (1993) Bird-habitat associations in the eastern Highlands. *Journal of Applied Ecology* **30:** 31-42.

The historic landscape of les bisses in the Valais, Switzerland.

D.S. Crook, C.O. Hunt and A.M. Jones

University of Huddersfield, Department of Geographical and Environmental Science, Queensgate, Huddersfield HD1 3DH, UK

Abstract

There are marginal landscapes in Europe that would not exist in their present form if earlier populations had not developed technologies and strategies to manage the environment in a sustainable manner. One such area is the Valais canton of Switzerland, where, without the intricate water management system of the *bisses*, the density of population and animals could not have been maintained. The *bisses* have helped create the landscape that is today so important to the tourist industry. This paper describes the technology and social organisation that have helped to perpetuate the *bisse* system since the twelfth century. Inheritance strategies and division of water rights are discussed in relation to their roles in creating a fragmented landscape. The problems of fragmentation are described and two examples of land and water redistribution and consolidation are presented.

Introduction

This paper deals with the consequences for an environmentally marginal mountain landscape of a sustainable long-term land use strategy. The *bisses* of the Valais canton, Switzerland, are a sophisticated response by the local population to the problem of aridity. This water management system provided the foundation for a distinctive agricultural style that has been practised since at least the twelfth century. The *bisse* system has maintained a density of population and livestock unachievable in a marginal environment without effective water and land management. The way in which water and land has been divided as a result of inheritance strategies and the division of water rights has led to the creation of a distinct fragmented landscape. This now provides the backbone of the Swiss tourist industry which is crucial to all levels of the economy and as such needs to be protected. The fragmentation of landscape, however, has presented problems in some communes as the management of both land and water rights has become unfeasible both from an administrative and physical point of view. Family responses to land and water fragmentation have until recently proved adequate, although the scale of the problem in some communes has led to the introduction of measures to consolidate and redistribute land holdings and water rights.

Valais canton

Valais (fig 1) combines the type of high alpine landscape which is crucial to the tourist industry with the intensive agriculture and industry of the Rhöne Valley floor. These sectors rely on water resources which are limited because Valais lies within a ring of high alpine mountains and is thus partly in rain shadow, so that areas of the canton can be defined as semi-arid.

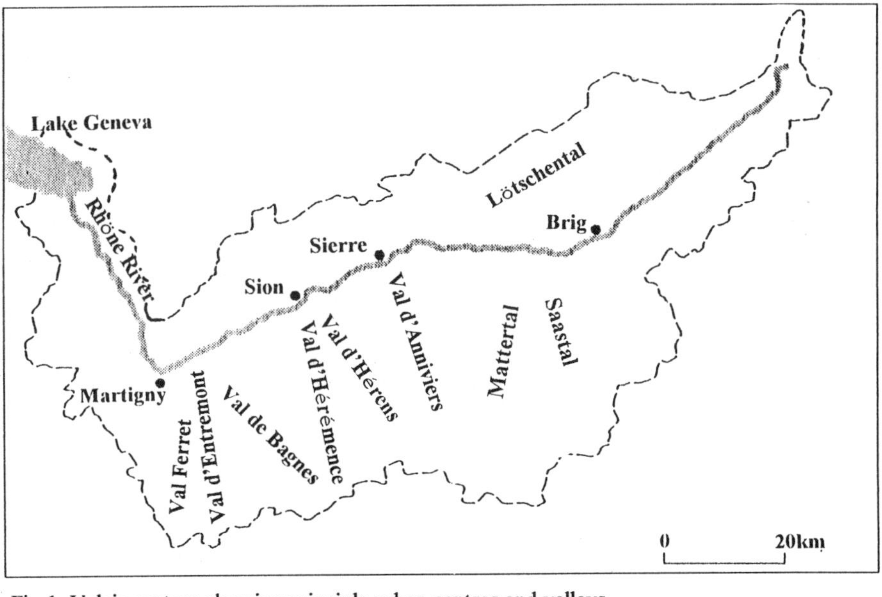

Fig 1 Valais canton: showing principle urban centres and valleys.

Annually the central Valais receives between 500 and 700mm of precipitation a lot of which falls as snow during the winter months (Aufdereggen & Werlen, 1993). Skies are generally clear, hours of sunshine long and average summer temperatures high. Precipitation increases with altitude such that the Pennine Alps and Bernese Oberland contain extensive glaciated areas. Meltwaters from winter snows and glaciers are and always have been crucial to the success of agriculture and industry in the area.

FRAGMENTATION IN AGRICULTURAL LANDSCAPES

Agriculture

The area available for agriculture and housing is restricted by the natural landscape and natural hazards. Steep and unstable slopes liable to slump, rockfall, seismic activity, floods, snow and ice avalanche all contribute to a highly active environment. Aspect is also crucial to the development of agriculture as winter snows thaw quicker from south facing slopes than those on north facing slopes. Climate and altitude gradients contribute to a diverse array of ecological zones. Strategies for alpine land use and land ownership have developed that clearly maximise the efficiency of management and use of resources. Transhumance took advantage of the seasonal expansion in the agricultural resource as cattle were moved up onto the alps when the spring thaw occurred. Alps and forests of the Valais have generally remained under communal ownership and management. Pastures, meadows, gardens, grain fields, villages and vineyards are normally owned and managed by individuals (Netting, 1972).

The use of terracing, a very distinctive feature in the central Valais, has been one response to steep slopes otherwise unfit for cultivation. By creating a terrace, the farmer effectively increases the surface area available for cultivation, increases the depth and moisture retaining properties of the soil and stabilises what might be, if left unmanaged, an unstable and potentially dangerous site. Terracing has allowed the growing of vines, a very profitable cash crop, and increased the efficiency and distribution uniformity of irrigation. Those areas close to and on the Rhöne valley floor are able to remain more economically competitive than the upper reaches of the lateral valleys of the Pennine Alps as they generally have a warmer, drier and sunnier climate allowing vines and fruit crops a longer growing season. Paradoxically, it is these areas which are most in need of irrigation since water resources are scant either as a result of the climatic conditions or competing uses for water. The Swiss, unlike most wine growing nations, choose to irrigate their vines to increase the size of the yield.

Irrigation

Without irrigation much of Valais would be a bare and barren landscape where only drought tolerant crops could survive. There have been a number of studies in the area to assess the water needs of different crops during the growing season. Muller (1946) found that the water needs during the growing season for hay meadows were approximately 10,000-12,000m^3/ha. Low annual rainfall and high rates of evapotranspiration lead to a water deficit, particularly in June, July and August, of 7,000-9,000m^3/ha for a hay meadow and 1,500-2,000m^3/ha for a vineyard. In dry years it may be necessary to start irrigating at the beginning of April and continue through to October. Depending on the geographical location and weather conditions during the agricultural year, irrigation can either allow crops to be grown in areas otherwise incapable of supporting them or increase the number, variety and quality of yields. The use of *bisse* irrigation has allowed a higher density of population and livestock.

Bisses

The early inhabitants of the Valais, like those in other semi arid mountainous areas, developed a system of water management that allowed the settlement of people and the development of sedentary agriculture. This water management system is commonly known as the *bisse* system in the French speaking Valais and *suonen* in the German area. *Bisse* is a name used to describe water conduits that bring water from higher altitudes to areas in need of irrigation. The term is also often used to describe the system of secondary and tertiary channels associated with the division of water and traditional methods of distribution. One of the earliest traces of a *bisse* dates back to the twelfth century (Ladurie, 1972) while the first documentary evidence relates to the thirteenth century (Rauchenstein, 1908). The occasional construction of new *bisse* continued up to the start of this century. The number of principal *bisse* (<1000m) has fallen from 207 with a total length of 1400km at the turn of the century (Rauchenstein, 1908) to 190 in 1993 with a total length of 760km (Aufdereggen & Werlen, 1993). Whereas all *bisses* were used for agriculture in 1908 now only 80% are used.

The vast majority of *bisses* rely on glacial meltwaters for their supply (Vautier, 1942). Water is diverted using a dam from a mountain stream into an intake structure, normally with sluices and an overflow, to control the quantity of water entering the *bisse*. Original channels were earth cut, wooden or cut into rock using picks. Stone was used to line and reinforce channels where needed. Sediments carried in glacial meltwaters played a crucial role in lining the bottom of many permeable earth cut and stone cut channels thus preventing excess infiltration. Most of the materials used in the construction of ancient *bisse* were supplied from local sources (Mariétan, 1948). Larch and pines from sustainably managed communal forests were used for wooden channels and supporting materials which were used where the ground would not permit the use of an open earth cut channel or where vertical rock faces had to be traversed. Metal conduits and concrete and plastic piping are now more likely to be used in the maintenance of old channels. Since 1908 many principal *bisses* have been covered or use tunnels to avoid danger spots, shorten water delivery times and reduce maintenance and evaporation.

<u>Bisse networks</u>

As the demand for water in any one area could not normally be met by a single *bisse* anything up to five *bisses* were constructed at different altitudes along a slope. Sometimes water was captured from different watersheds. For example, the *Bisse de Saxon,* which is 32km long, traverses through three other communes before arriving at the Mayens of Saxon. Many of these more spectacular *bisse* have had a considerable impact on the landscape. In particular, the *Grand Bisse de Lens* and the *Bisse de Sillonin,* between Sion and Sierre, are clearly visible from the valley floor and just beyond Sierre can be seen the *Grosswasserleitung* and the *Mengiwasser* both taking waters from the Raspille river.

There is also a network of smaller *bisses* owned by individuals or small groups. These may be on the scale of a small spring being channelled to irrigate a single field or part of a field. This smaller network of *bisses* coupled with the system of secondary and tertiary channels and a number of small storage reservoirs have in many areas had a considerable impact on the landscape. In some areas, such as Les Haudères in the Val d'Hérens, traditional irrigation practices, now abandoned, can be identified by the landscape they have helped create. Tiny agricultural plots, often no more than $20m^2$, bordered on all sides by distributary channels, have developed a distinct reticulation pattern as a result of accreted sediments. The distribution network has in many areas served also as an indicator of property demarcation.

The distribution network

A *bisse* can often undergo a number of divisions or *répartitions*. This *répartition* is achieved using simple technology such as a dam (*barrage*) or a sluice *(écluse)* which can be made of wood, stone or metal. The responsibility for *répartition* varies from *bisse* to *bisse.* It is, however, normal to find either a *garde* or *répartiteur* who will oversee and monitor the action. In general there are two main systems of distribution onto the fields, the traditional method known as *ruissellement* and the modern method known as *aspersion* or spray irrigation.

Ruissellement

This method relies on the simple blocking and diversion of flow along a secondary or tertiary channel with a piece of wood, metal or stone *(étanche)*. Alternatively a sod of earth may be cut in the channel sides using a special tool known as a *delabre,* thus breaching the channel temporarily. Water then washes through the side of the channel and flows under gravity down slope. This will continue until the required ground is flooded. At this point the sluice is removed and placed further along the distributary channel allowing the irrigation of another section of the field. There has been a decline in the use of *ruissellement* since the turn of the century. This has led to a reduction in the number of secondary and tertiary distributary channels.

Aspersion

This method relies on *bisse* as a system of transporting water to either communal reservoirs (*etangs*) or smaller private storage tanks and cisterns. From these reservoirs water is brought under pressure, usually via hose pipes, where it is emitted from a rotational tap. Since the turn of the century there has been an increase in the use of spray irrigation.

Consortages

The construction of larger *bisse* was normally paid for by groups of wealthy individuals from a commune or by individuals from two or more communes, who formed themselves into an independently organised *consortage* (Aufdereggen & Werlen, 1993). In other cases the commune would help to construct and maintain the bisse and organise the administration. The *bisse consortages* were and still are essentially self-regulating, self-organised bodies. Each *consortage* or commune has a *statute* and *protocol* that sets out the system of control, division of water rights and responsibility for maintenance and cleaning. The *consortage* committee will generally act in the interests of the *consorts* taking major decisions about work needed, plans to increase the water supply and ameliorate any disputes between *consorts.* If disputes cannot be settled or someone constantly breaks the rules, they may be liable to a fine or referred to the communal court where they can be prosecuted under civil law. In some places *consortages* have relinquished their rights to water as the number of people using the water for irrigation for agriculture has diminished. In some areas the commune has bought or been given these rights to water and have assumed control of the *bisse.* Where the agricultural role of *bisse* is still important the system of regulation essentially remains the same.

Water Rights

Each *bisse* has its waters divided into units (*bulletins)* according to the number of people in the consortage or commune (Vautier, 1942). Water rights (*droits d'eau)* are then attached to these *bulletins.* Normally, people would receive water rights proportional to the amount of land owned and contribute proportionately to the building and maintenance of a *bisse.* However, water rights are not always attached to the land and in some cases can be sold, exchanged or loaned independent of land. A water right equates to a volume of water over a given period of time and has a monetary value. Depending on the size of the *bisse* and the number of communes utilising the *bisse,* there may be a number of divisions and subdivisions (*répartitions*). In turn there may be a complex system of rotations (*tours*) relating to a set time period (a number of days) and possibly different *répartitions.* The *tours* will commence from the date of opening of the system and finish after the last *droit* is taken from the final *tour.* The system of *répartitions, tours and droits* was flexible enough to allow for expansion in the population and increasing agricultural demand for water.

Fragmentation of land and water rights

Fragmentation of land was an inevitable consequence of a system of partible inheritance in which each family member received an equal inheritance. The fragmentation of water rights was a logical response. At first a family may possess a number of *droits* relating to a number of days, hours, rotations and possibly different *répartitions.* An equitable distribution of these *droits* would result up to the point where a whole division could be made. When this was no longer possible *droits* were often split into fractions. At Visperterminen, at the head of the Saastal valley, a *droit* was referred to as a *quart* and was equivalent to four hours

(Vautier, 1942). This *quart* could be divided into halves (2 hours) quarters (1 hour) and eighths (1/2 hour). In some places further fragmentation has led to the inheritance of fractions of *droits* equivalent to 15 minutes. The ability to inherit, buy, exchange and loan *droits* in many *consortages* and communes has meant that farmers have had the opportunity to acquire more rights or relinquish these rights. Almost inevitably in some areas the fragmentation of water rights and land has presented problems as it simply becomes inefficient to water such a short amount of time or to manage such small plots. Alternative income opportunities have been made possible by the improved transport access to valleys. This has meant that inheritors of land and water are able to form family agreements as to who manages the land and how it is managed. It is common practice for many farmers to be part time and earn dual incomes. This method of self regulation has until recent times been effective. In some areas, however, other measures have been needed.

Consolidation

Changes to the *statute* of the *Consortage* of the *Bisse de Salin* in 1961 (art. 6) sought to rectify the problem of fragmentation of water rights. The *consortage* decided to make the smallest allowable division half an hour which is equal to half a right. Any *consorts* with smaller divisions were given one year to modify their fraction. If at the end of this period the division was less than an hour, the right was cancelled. An indemnity could be paid of one franc a minute to make the time up to a half hour. In addition only one member of a family was allowed to represent them at meetings of the *consortage*. This process appears to have been designed to simplify the administration of the *consortage* but would also have served to consolidate land holdings.

Grove and Grove (1990) noted that the commune of Ayent was the first in the Valais to attempt land redistribution and consolidation. In a process started in April 1930 the proprietors of the Mayens of Anzere (extent 175ha) constructed 6km of road and 6km of secondary tracks. The number of holdings was then reduced from 2768 to 350 (counting husband and wives as a single proprietor) with the average proprietor having 1.2 holdings. Prior to this redistribution the average size of holding (counting husbands and wives separately) was only 632m^3 with each proprietor having an average of 8 holdings. In the 1960s arable land on lower slopes close to the villages was redistributed so that 10000 holdings were consolidated into about 1000 holdings with one holding going to each proprietor. This process also included the building of new roads and the relaying of distributary *bisse* often in concrete or using pipes. The federal government met 50% of the costs because of the commune's intentions to consolidate their holdings. In addition the canton subsidised 30% of the costs with the final 20% being evenly split between the commune and proprietors.

A possible future for the bisses?

The process of modernisation and the general decline in the number of *bisses* paradoxically threatens the landscape that is important to the tourist industry. The cantonal authorities and a number of communes have made moves to realise the tourist, agricultural, cultural and landscape potential of the principal *bisses*. The canton have listed the current status, abandoned or functioning, and establishing the present, historic, agriculture, tourist and landscape importance of all principal *bisses*. From this assessment a *bisse* is classified as either locally, regionally or cantonally important. Principal *bisses*, in general, make excellent footpaths because of their gentle slopes. Over half of those remaining are pedestrianised. There are schemes currently underway to regenerate *bisse* both selectively in sections such as the Riddes section of the *Bisse de Saxon* and in their entirety such as the *Grand Bisse de Vex*. The former solely for tourism and the latter for a mixture of tourism and agriculture. Both of these *bisse* play an important security role during dry periods in providing water for garden irrigation which normally relies on drinking water supplies. This need to reduce the demand on drinking water supplies and the possibility of climatic and economic change in the future may lead to a resurgence in the agricultural importance of the *bisses*. The mechanics of the system have proved to be robust enough to have withstood environmental, technological, social and demographic change over a period of at least 800 years. As such the *bisses* of Valais may provide a blueprint for similar sustainable systems.

References

Aufdereggen, J. & Werlen, C (1993) *Rapport bisses/suonen*, Canton du Valais, Département de l'environment et de l'aménagement du territoire, unpublished.

Grove, A.T. & Grove, J.M. (1990) Traditional montane irrigation systems in modern Europe: an example from Valais, Switzerland. *Agriculture, Ecosystems and Environment* **33:** 181-186.

Ladurie, E. Le R. (1972) *Times of feast times of famine: a history of climate since the year 1000*, G. Allen & Unwin, London.

Mariétan, I. (1948) *Les bisses: La lutte pour l'eau en Valais*, Editions du Griffon, Neuchatel.

Muller, H. (1946) De quelques solutions nouvelles du probleme de l'irrigation. *Bulletin Murithienne* **63:** 33-40.

Netting, R. McC. (1972) Of men and meadows: strategies of alpine land use. *Anthropological Quarterly* **45:** 132-144.

Rauchenstein, F. (1908) *Les bisses du canton du Valais*, Départment de l'Intérieur, Switzerland.

Vautier, A. (1942) *Au pays des bisses*, Lausanne.

Pond landscapes: fragmentation pressures & survival mechanisms

J. Boothby, A.P. Hull, & D.A. Jeffreys

Ponds Research Unit, Liverpool John Moores University, Trueman Building, Webster St, Liverpool L3 2ET, UK

Abstract

The ponds of Cheshire have declined in number from the middle of last century. Created mainly by marl extraction, their loss is attributable *inter alia* to changed farming practices, vegetational succession, and various counter-urbanisation forces. Fragmentation of the farmland pondscape has resulted. With little available legal protection for ponds, tackling degeneration and loss might best be undertaken by community networking. Information on current pond status can be used as the basis for an informed pondscape recovery strategy.

Introduction

With some 16,000 farmland ponds, the county of Cheshire can justifiably be considered the 'pond-capital' of Britain. So densely is the county populated with such water bodies (Figure 1) that the term 'pondscape' seems appropriate. The origin, present status, and uncertain future of the pondscape are investigated, and a mechanism for combatting fragmentation proposed.

The Cheshire pondscape

Origins of ponds

Though some water bodies in Cheshire have resulted from glacial and periglacial processes (25,000bp - 13,500bp), the origin of the majority is anthropogenic, corroborated by evidence from the 1:1250 and 1:2500 First Edition Ordnance Survey mapping (1872-1880) such as a distinctly angular shape and information initially collected in the field by Parish surveyors (Harley, 1979). The apparent origins of the bulk of ponds are sand or gravel workings, brick-clay extraction, and, most importantly, marl-pit excavation.

Marl excavation was undertaken for top-dressing the pastureland and was displaced as a practice when more effective fertilizers (eg: crushed bone; guano; nitrates) became available. By 1830, the creation of small marl pits had ceased to be economic: some large-scale mechanized works were however undertaken in parts of Delamere Forest in the 1860s (Grantham, 1864). Following marl removal, the ponds were available for

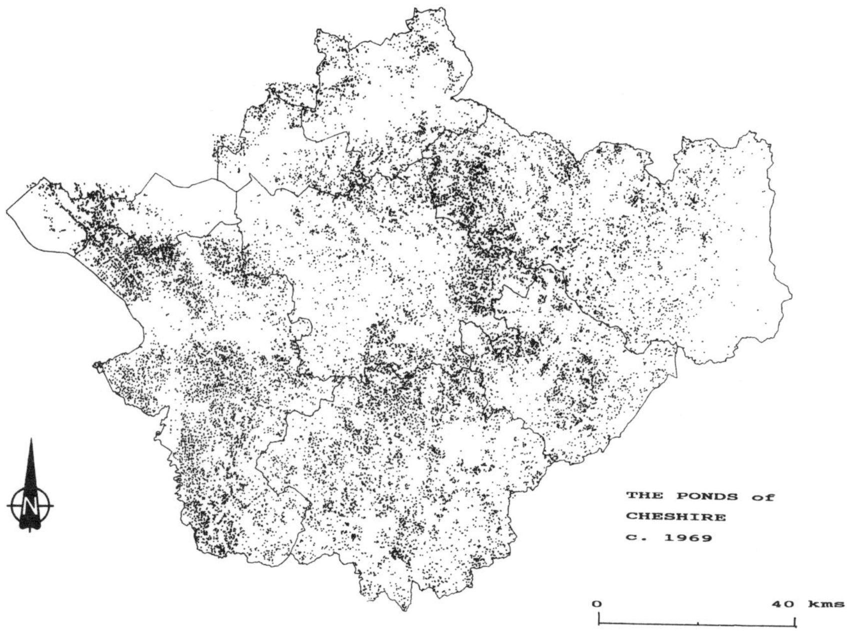

Figure 1: The ponds of Cheshire, c. 1969: Some 23,000 ponds are shown here; compiled by authors.

---

stock watering and other farm water needs; some ponds were also created specifically for these latter purposes.

Size, shape, quality, location

Most of the 42,000 ponds recorded on the early (1870s) large scale OS maps are small and angular. Those that remain today have typically decreased in angularity with vegetational succession, and many ponds have significant copse development around them; this speeds desiccation and the associated leaf litter may help to produce anaerobic conditions in some ponds.

Table 1 Pond:soil association, Cheshire (Source: authors)

Soil type	Area km² (a)	Area %total (b)	Ponds n (c)	Ponds %total (d)	density (c / a)
Gley Soils					
Raw	42	1.8	33	0.1	0.8
Surface water	1225	52.4	17669	77.3	14.4
Ground water	243	10.4	1455	6.4	6.0
Peat soils	27	0.0	103	0.5	3.8
Brown soils	500	21.4	2667	11.7	5.3
Podzolic soils	77	3.3	158	0.7	2.1
Lithomorphic soils	2	0.1	nil	nil	0.0
Urban areas	218	9.3	722	3.4	3.5
Total	2337	100.0	22857	100.0	9.8

Note: Pond locations digitized from OS 1:10,000 @ 1969 -/+ 5yrs

The widespread presence of *Typha latifolia* (Reedmace) and *Phragmites australis* (Common Reed) has significantly reduced the area of open water in many ponds, except where counteracted by management for angling or continued trampling by cattle. Stratified surveys (n=60) across the county show a median pond length (major axis) of 25.7 metres. In vertical cross-section, ponds typically show an asymmetric profile; though some deep ponds are to be found, most are little deeper than 2 metres. As Table 1 demonstrates, ponds (of all origins) show a clear and significant association with soil-type; they are at greatest densities on the seasonally wet ground water gley soils and least common on the freely draining podzolic soils.

Though perhaps fewer than 5% of sample ponds show evidence of gross pollution (Boothby *et al.*, forthcoming) there are reasons to believe that nutrient availability is fairly high, probably from fertilizers leached out from surrounding soils and transported via ground- and ditchline-drainage which now connects many of the ponds. Given the ground conditions in Cheshire, it is not surprising that pond-water pH levels are fairly high: the median pH value is 7.285 and a maximum of 9.44 has been recorded. Such waters support 5 of the 6 British amphibian species; various surveys (Boothby *et al.*, 1993; Anderson, personal communication) suggest that more than one-third of ponds have *Rana temporaria* (Common Frog) present, 10% have *Bufo bufo* (Common toad), 25% have *Triturus vulgaris* (Common or Smooth newt), and just under 10% have *T. cristatus* (Great Crested newt).

Pond loss

Scale and cause of pond loss

Pond loss from 1872 to 1969 was substantial, on average under 0.5% per year. But throughout the post-war years, loss rates have been much greater, approaching 2% annually, to reach a 1989 total of 17,000. Table 2 shows that such high loss-rates are not uncommon. Projections based on these rates suggest that,

Table 2 Pond Loss: some comparative rates

	Annual loss (%) between: c 1870		c.1969		1985		1992
Cheshire County[a]	x	0.5	x	0.9	x		
Warrington Dist.[a]	x	0.6	x	1.1	x		
Chester Dist.[a]	x	0.4	x	0.9	x		
Tarporley Parish[a]	x	0.6	x		1.9		x
Lancashire County[b]			x		>1.3		x
Ormskirk Parish[c]	x	0.7	x		0.9		x
Bolton MBC[d]			x		>1.3		x
Midlands counties[e]					x	>3.9	x
Gwent County[e]					x	>2.6	x
England and Wales[f]	x		0.5		x		

[a] computed by authors [b] Mullaney (1991) [c] McCartney (1993)
[d] Boase (1991) [e] Pond Conservation Working Group (1992)
[f] Rackham (1986)

by 2050, almost all Cheshire ponds below 2 hectares in area will have been lost (Boothby *et al.*, 1993). The effects of these losses on pond density have been quite marked: from an overall density of 11.9 ponds km^{-2} recorded from the 1872 survey, density had fallen to only 4.9 km^{-2} in 1985. Detailed tabulation of the causes of this loss are not yet complete, and are awaiting fuller interpretation of recent air photographic coverage. However, the following loss mechanisms are known to be significant: vegetational succession; ploughing-out and draining; solid waste disposal; development for industrial and commercial land; transport infrastructure; housing growth; recreational facility development, including golf-courses. Changes in those farming systems which brought about the creation of ponds have been responsible for their removal. There is, perhaps, no more banal epitaph to the farm pond than the plastic, mains-fed cattle trough at the field boundary. Post-war intensification of agriculture has impacted upon farming in many ways, all to the detriment of farmland ponds and their proximate habitats; these impacts continue (Barr *et al.*, 1994).

Loss of pondscape

The consequences of pondscape loss in Cheshire have not yet been fully quantified, but logically include (i) habitat loss resulting from pond removal and pond degradation; (ii) increased fragmentation of the pond patches; and (iii) amenity loss with both visual and social expressions.

Habitat loss and fragmentation

In addition to losses of individual ponds, habitat loss is occasioned by pondscape fragmentation; for mobile species, the effects are felt beyond the individual pond. Gabb and Kitching (1991) comment upon the impact of pond loss on dragonfly populations in Cheshire:

> "Loss of individual ponds may impoverish the (dragonfly population) of a much wider area, because remaining healthy but scattered ponds may not be adequate to support viable populations of all species" (Gabb and Kitching, 1991; p.57)

In the Netherlands and Flanders, research on amphibians has stressed the importance of pond availability within a species' home-range. Many species make use of a variety of ponds over a number of years: it is thus important that sufficient ponds exist to enable species persistence within an area (Laan and Verboom, 1990). Clearly, pond clusters are especially valuable for this. The critical distance to the nearest-neighbour pond seems to vary by species: for newts, this distance appears to be about 500m; for certain invertebrates, 750m; and about 1000m for frogs and toads (Furniss and Lane, 1992; Laan and Verboom, 1990).

But equally important is the nature of the space between ponds. For example, amphibia, which have both an aquatic and terrestrial phase, require buffer zones around, and corridors between, habitat patches. In an ideal pondscape, this interstitial land would contain deciduous woodland, scrub, and rough grassland. In planning terms, Vos (1993) and Opdam *et al* (1993) have demonstrated how these elements may be modelled and used to suggest optimal spatial strategies for reintroduction declining species such as *Hyla arborea* (Tree frog) in areas of the Netherlands.

Amenity loss

The visual and social expressions of amenity loss are significant, if rather more problematical to model. Water features are known to be consistently prominent in visual assessments of environmental quality (Dearden, 1986) and in visual amenity evaluations for recreational planning (Jubenville *et al.*, 1987); additionally, various researchers have placed emphasis on the significance of water as a significant component of landscape aesthetics (eg. Appleton, 1975). Current popular concern with environmental quality is quite high in Cheshire (Gittins, 1993). Moreover, ponds and their associated surrounding copse vegetation can occupy salient landscape positions in a county of low, rolling relief; they are typically valued for their contribution to landscape heterogeneity. Ponds also have specific recreational use-value as angling resources and for wildfowling amongst specific sections of the population. Ecologically, ponds also provide a valuable resource for education or for private study and enjoyment (Pond Action, 1993).

Combatting pond and pondscape loss

Attempts to combat pond loss are surrounded by serious difficulties. Ponds have no intrinsic legal protection, though certain species have protection inferred by, *inter alia*, the Wildlife and Countryside Act, 1981 and the Water Act, 1989. Nor is scientific protection really a match for legitimate private farming interests or for benign neglect. And though agricultural policy has the potential to encompass pond (and other habitat) protection within a more meaningful Set-Aside policy, this is not yet taking place.

It is fortunate, then, that in policy terms Cheshire County Council has already begun to address some of the issues in pond landscape change. The County Structure Plan already specifies a general policy on development and landscape:

> "Development proposals which involve the loss of ponds, wetlands, heathlands, ancient grasslands or ancient woodlands will not normally be allowed and their conservation will be encouraged" (Policy: Landscape ENV12); Cheshire County Council (1990).

In practice, the Council operates a policy of 'no net loss' - a new-for-old approach to development proposals which helps to maintain a varied stock of ponds. Some of the District Councils within Cheshire have prepared pond management plans (eg. Commission for New Towns/ Warrington Borough Council, 1993). But on farmland, free from most constraints of planning legislation, the loss continues apace.

Without legal protection and with now limited use-value, we must seek other mechanisms for pond protection. Involvement of the community seems to be both a virtue and a necessity: individuals and the wider community often have an active concern for their immediate landscape and can express this concern in a non-bureaucratic manner amongst their peers. Accordingly, from November 1993 we have developed, with local authorities, FWAGs, and local nature conservation groups a scheme for Parish Pond Wardens who can perform a monitoring role for their local ponds and can, where access is allowed or can be negotiated, undertaken appropriate renovation tasks; the Wardens may be individuals, groups, or organisations (such as schools, parish councils, or nature conservation groups). At its most basic, such a scheme can help provide information to understand the local context of pondscape deterioration.

Developing a strategy

The strategic issues in pondscape conservation require a different perspective (Grayson, 1993). It is unrealistic to aim totally to prevent pond loss. But equally, we need to be able to identify the strategically optimal locations for intervention via pond management, conservation, or (re-)construction. Whilst the detailed recommendations for Cheshire are yet to be formulated, and must certainly contain arrangements for appropriate pond management, we can suggest 5 basic ground-rules:

i. sustain the maximum number of ponds: in the absence of other actions, this maximizes habitat potential;

ii. maintain a stock of ponds at various stages of vegetational succession: a new-for-old policy will assist this aim;

iii. all things being equal, ensure ponds are sustained in clusters within a maximum of 500m of each other; where possible, aim to link clusters together;

iv. ensure that the inter-patch areas contain appropriate habitats to facilitate species movement and use;

v. aim to link ecologically valuable but isolated ponds into the larger pondscape.

References

Appleton, J. (1975) *The experience of landscape*, Wiley, London.

Barr, C.; Benefield, C., & Howard, D. (1994) *Changes in the number of inland water bodies in Britain between 1984 and 1990*, ITE, Merlewood.

Boase, M. (1991) Disappearing ponds. *Lapwing*, Winter, 16-17.

Boothby, J.; Hull, A. P., Jeffreys, D.A., & Small, R.W. (forthcoming) *Wetland loss in North-west England: the conservation and management of ponds in Cheshire*, in Hughes, J. and Heathwaite, A.L.: *Hydrology and hydrochemistry of wetlands*, Wiley, London (Chapter 27).

Cheshire County Council (1990) *County Structure Plan Review*, CCC, Chester.

Commission for New Towns/ Warrington Borough Council (1993) *Warrington ponds: Management Plan,* Gillespies Consultants, Warrington.

Dearden, P. (1986) Consensus and a theoretical framework for landscape evaluation, *Journal of Environmental Management* **24:**

267-278.

Furniss, P. & Lane, W. (1992) *Practical conservation: Water and wetlands*, Hodder & Stoughton, London.

Gabb, R. & Kitching, D. (1991) *The dragonflies and damselflies of Cheshire*, National Museums and Galleries on Merseyside, Liverpool.

Gittins, J. (1993) *Community involvement in environment and recreation*, **in: Glyptis, S.** (Ed): *Leisure and the environment*, Belhaven, London, 183-192.

Grantham, R.B. (1864) A description of the works for reclaiming and marling parts of the late Forest of Delamere, *Journal of the Royal Agricultural Society of England* **1 (25):** 369-380.

Grayson, R.F. (1993) The distribution and conservation of the ponds in North-west England, *Lancashire Wildlife Journal* **2 & 3:** 23-51.

Harley, J.B. (1979) *The Ordnance Survey and land-use mapping: Parish 'Books of Reference' and the County Series 1:2500 maps, 1855-1918*, Historical Geography Research Group/ GeoAbstracts, Norwich.

Jubenville, A.; Twight, B.W., & Becker, R.H. (1987) *Outdoor recreation management: theory and application; Chapter 11: Visual resource management*, revised edition, Venture Publishing, Penna.

Laan, R. & Verboom, B. (1990) Effects of pool size and isolation on amphibian communities, *Biological Conservation* **54:** 251-262.

McCartney, I. (1993) *Pond decline in Lancashire*. Unpublished dissertation, Liverpool John Moores University.

Mullaney, A. (1991) Marl pits - Lancashire's precious pearls. *Lapwing*, Winter, 12-13.

Opdam, P.; van Apeldoorn, R. Schotman, A. & Kalkhoven, J. (1993) *Population responses to landscapes fragmentation*, **in: Vos, C.C. & Opdam, P.** (Eds) *Landscape ecology of a stressed environment*, Chapman & Hall, London, 147-171.

Pond Action (1993) *A future for Britain's ponds: an agenda for action*, Pond Conservation Working Group, Oxford.

Pond Conservation Working Group (1992) *Protecting Britain's ponds (Draft)*, Oxford, PCWG; (mimeo).

Rackham, O. (1986) *The history of the countryside*, Dent, London.

Vos, C. C. (1993) *Versnippering en landinrichting in Zeeuws-Vlaanderen: Deel I - Boomkikkers*, Instituut voor Bos- en Natuuronderzoek, Wageningen.

Grassland Diversity in Relation to Field Parcel Size and Management

Alan Cooper, Thomas McCann and James Power

Department of Environmental Studies, University of Ulster,
Coleraine, Northern Ireland, BT52 1SA

Abstract

A plant ecological survey of lowland grassland contributes to more informative models of the effects of fragmentation on species-rich grassland. The structured sampling programme provides a framework for defining the composition of species-rich grassland and for estimating differences in the area of land occupied by contrasting grassland types. This allows formal definition of habitat size and quality and their spatial and temporal variation.

The main types of species-rich grassland in lowland landscapes were highly fragmented. Species diversity did not decrease with field parcel size or in part-field parcels compared with full-field parcels. The main conclusion is that geographical factors associated with fragmentation have a relatively unimportant influence on diversity compared with spatial and temporal variation in grassland management for agriculture.

Introduction

The Northern Ireland Countryside Survey (NICS) recorded the land cover composition of 628, 25ha sample grid squares between 1986 and 1991 (Murray *et al.* 1992). Analysis of the resulting map archive and computer database established estimates of the area of the main types of woodland, semi-natural vegetation, agricultural grassland and crops in NI and its regional landscapes. The sampling programme was based on a multivariate NI Land Classification (NILC) which acted as a sample stratification for field work and defined landscape types (Cooper 1986). The botanical composition of eight grassland land cover types, mapped for the NICS, was subsequently determined by Cooper and McCann (1994).

Grassland Land Cover Types

There were three types of agricultural grassland land cover (ryegrass, mixed species grassland and other agricultural grassland). They were dominated by agriculturally preferred species and had a relatively low species diversity (Table 1). There were five seminatural grassland land cover types (species-rich dry and wet grassland, fen meadow, calcareous grassland and hill pasture). These were largely unimproved and had a higher species diversity than the more intensively managed agricultural grassland types.

Ryegrass grassland: intensively managed grassland with *Lolium perenne.*

Mixed species grassland: longer established, grazed swards on drier soils. A high frequency and cover of *Cynosurus cristatus* is characteristic. Other species with a relatively high frequency are *Agrostis capillaris, Anthoxanthum odoratum, Bellis perennis, Plantago lanceolata, Prunella vulgaris, Rhytidiadelphus squarrosus, Rumex acetosa* and *Trifolium repens.*

Other agricultural grassland: long-established grassland, commonly on less well drained or peaty soils. It includes degraded swards formerly more productive species-rich grasslands now intensively managed, and neglected swards. Species such as *Agrostis* spp., *Alopecurus pratensis, Elymus repens, Holcus lanatus* and *Ranunculus repens* are often dominant. Infestation with *Juncus effusus* or *Juncus articulatus/acutiflorus* is frequent.

Species-rich dry grassland: unimproved on moderately well drained mineral soils. *A. capillaris* and *Festuca rubra* are major constituents, often with much *C. cristatus* and *P. lanceolata.* Characteristic forbs are *Centaurea nigra, Hypochoeris radicata, Lathyrus pratensis, Lotus corniculatus, Ranunculus bulbosus, Rhinanthus minor* and *Trifolium pratense.*

Species-rich wet grassland: unimproved, poorly drained, with *J. articulatus/acutiflorus* and *J. effusus* generally abundant. *Agrostis stolonifera, Carex nigra* and *Poa trivialis* can achieve high cover values. Other characteristic species are *Cardamine pratensis, Equisetum fluviatile, Equisetum palustre, Filipendula ulmaria, Galium palustre, Ranunculus flammula, Senecio aquaticus* and *Stelleria alsine.*

Fen meadow: occurs on poorly drained mineral soils often with a thin peaty layer. Characteristic species are *Agrostis canina, Cirsium dissectum, Carex panicea, Carex* spp., and *Succisa pratensis.*

Calcareous grassland: on thin soils over chalk or limestone. Characteristic species are *Alchemilla vulgaris, Carex flacca, Galium verum, Hieracium pilosella, Koeleria macrantha, Linum catharticum, Sesleria albicans* and *Thymus praecox.*

Hill pasture: occurs on moderately well drained mineral or peaty soils. *A. capillaris* and *F. rubra* dominate the sward.

Grassland Sampling Programme

A stratified random sample of 498 grassland field parcels was located in a random sub-sample of 190, 25ha grid squares taken from the NICS database. The sample was stratified by grassland land cover type, multivariate land class and field parcel size-class. Grassland parcels < 0.05ha were not sampled. Approximately 50 sample units (field parcels) were selected from each of the grassland types. The number of field parcels from species-rich dry and wet grassland was subsequently increased to include more large field parcels.

Changes in land cover type since baseline survey, checked against standard land cover descriptors, were recorded. Within each field parcel, one 4m^2 quadrat nested within a random 200m^2 quadrat was located to record species composition. A Domin cover value was assigned to each species in the 4m^2 quadrat and the presence of species was recorded in the 200m^2 quadrat.

Grassland Classification

A TWINSPAN classification (Hill 1979) of the 200m^2 quadrats was carried out. Species occurring in three or less quadrats were omitted from analysis. The classification hierarchy was stopped using group size and the ecology of the TWINSPAN indicator species. Seven vegetation groups were produced by the classification (Table 1). There was a strong correlation between land cover type and vegetation group. A canonical discriminant function analysis (SPSS, 1990) was carried out on the $4m^2$ quadrat data to identify the species by which they could be separated. Each quadrat was also classified by the discriminant analysis, to a land cover type (Table 2). Overall, the discriminant function analysis assigned 90% of quadrats to the same land cover type as the field parcel from which the quadrat was recorded. The results show that the land cover mapping procedure used for the NICS has a clear ecological basis, with each land cover type being associated with a distinct species assemblage.

Table 1. The number of $200m^2$ quadrats in TWINSPAN groups and land cover types. Group names represent species with the greatest cover and frequency.

	Land Cover Type							
	Agricultural grassland			Seminatural grasland				
TWINSPAN group	1	2	3	4	5	6	7	8
Lolium perenne/ Poa trivialis	55	11	25	-	-	-	-	-
Holcus lanatus/ Ranunculus repens	2	25	19	11	28	8	-	2
Holcus lanatus/ Festuca rubra	-	2	1	15	17	-	-	1
Anthoxanthum/ Agrostis capillaris	-	1	6	-	15	9	2	-
Potentilla erecta/ Anthoxanthum	-	-	-	14	4	7	21	-
Holcus lanatus/ Juncus effusus	-	2	7	1	1	53	13	-
Festuca rubra/ Prunella vulgaris	-	-	-	1	1	-	-	41
No. of samples	57	41	58	42	66	77	36	44
Species density (standard error)	14.5 (0.7)	25.0 (0.9)	22.8 (1.1)	32.3 (1.5)	34.7 (1.4)	34.0 (1.1)	37.3 (1.7)	53.1 (1.8)

Land cover types are: 1. Ryegrass pasture; 2. Mixed species grassland; 3. Other agricultural grassland; 4. Hill pasture; 5. Species-rich dry grassland; 6. Species-rich wet grassland; 7. Fen meadow; 8. Calcareous grassland.

Land Cover Estimates

Estimates of the area of grasslands in NI (Table 3) were calculated by database programming (Murray *et al.* 1992). Ryegrass, other agricultural grassland and mixed species grassland comprised 95.5% of the grassland resource in lowland landscape. Twice as much seminatural grassland was found in the uplands. Calcareous grassland and hill pasture, much of which was unenclosed, were both upland resources. Species-rich wet and dry grasslands and fen meadow made up most of the 4.5% seminatural grassland in lowland landscapes.

Table 2. The percentage of $4m^2$ quadrats assigned to a land cover type by discriminant function analysis.

Predicted Land Cover Type	Quadrats							
	1	2	3	4	5	6	7	8
1. Ryegrass pasture	88	2	12	-	-	-	-	-
2. Mixed species grassland	5	93	4	3	1	10	2	-
3. Other agricultural grassland	7	5	81	3	1	5	2	-
4. Species-rich dry grassland	-	-	-	91	1	-	-	6
5. Species-rich wet grassland	-	-	4	2	95	-	-	-
6. Hill pasture	-	-	-	2	-	86	-	-
7. Calcareous grassland	-	-	-	-	-	-	96	-
8. Fen meadow	-	-	-	-	1	-	-	94

Land cover types are: 1. Ryegrass pasture; 2. Mixed species grassland; 3. Other agricultural grassland; 4. Hill pasture; 5. Species-rich dry grassland; 6. Species-rich wet grassland; 7. Fen meadow; 8. Calcareous grassland.

Table 3. Estimated mean percentage grassland area with standard error.

Land Cover Type	Lowland		Upland		N. Ireland	
Calcareous grassland	0.03	(0.02)	0.19	(0.05)	0.08	(0.02)
Species-rich dry grassland	0.56	(0.10)	0.41	(0.12)	0.53	(0.08)
Species-rich wet grassland	1.47	(0.18)	1.60	(0.27)	1.52	(0.15)
Fen meadow	0.38	(0.09)	0.50	(0.15)	0.42	(0.08)
Hill pasture	0.12	(0.03)	2.40	(0.54)	0.87	(0.18)
Ryegrass pasture	37.20	(1.60)	14.30	(1.70)	29.70	(1.23)
Mixed species grassland	3.40	(0.50)	2.90	(0.40)	3.30	(0.33)
Other agricultural grassland	26.50	(1.40)	16.60	(1.80)	23.20	(1.16)

Lowland = NI land classes groups 1-16. Upland = NI land class groups 17-23.

Species-area Relationships

The species diversity (mean number of species per $200m^2$ quadrat) of full-field parcels (areas of land management demarcated by field boundaries) and part-field parcels (areas of vegetation demarcated by changes in land cover-type within field parcels) is plotted for species-rich dry grassland in Fig. 1. There was a negative relationship between species diversity and parcel size. The larger parcels had a lower diversity. This relationship was similar for part-field parcels and full-field parcels. The diversity of small parcels (< 0.5ha) was variable; some with a high diversity, others with a relatively low diversity, thus indicating a high degree of inter-parcel heterogeneity. The same relationships held for species-rich wet grassland and fen meadow.

In order to determine if there was any difference in the types of species associated with different sized parcels, species were partitioned into two groups representing those considered to be characteristic of Irish mestrophic grassland and nitrogen indicator species (*sensu* Ellenberg 1974). There was a greater number of both mesotrophic grassland and nitrogen indicator species in the smaller parcels of species-rich dry grassland. The greater numbers of nitrogen indicator species in the smaller parcels does not reflect any fundamental difference in soil nitrogen status, however, since the greater the number of species there were in a quadrat, the more nitrogen indicator species were recorded (Fig. 2).

Fig. 1. The number of plant species recorded against parcel area.

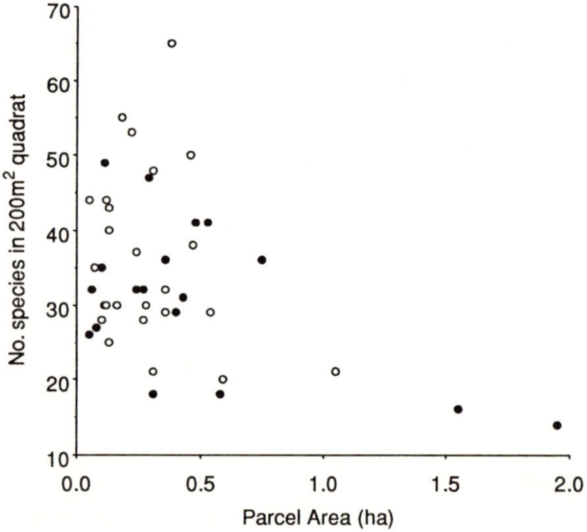

o = part-parcel; • = field parcel.

Grassland Management

A minimum variance cluster analysis (Wishart 1978) was used to classify the 4m^2 quadrats into vegetation communities. Fifteen groups of quadrats were delimited by stopping the classification at an agglomeration coefficient of 40. At this level there was a high degree of dissimilarity between groups and below it, small clusters of quadrats were chained off. Ordinations of the 15 end groups indicated no marked composition gradients across which the quadrats might have been split further. Grassland vegetation communities with the lowest mean species densities had the greatest numbers of nitrogen indicator species (Fig. 3). These vegetation types included agricultural grassland dominated by ryegrass and grassland defined as neglected (ungrazed grassland with a high biomass of accumulated litter).

Fig. 2. Species rich dry grassland; number of species per 200m² quadrat against mean number of nitrogen indicator species.

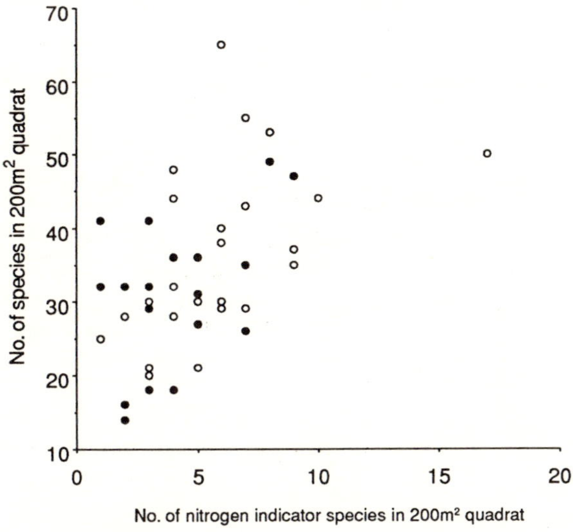

o = part-field parcel; • = field parcel.

Land cover change

Detailed analysis of the rate of land cover change between baseline survey and monitoring resurvey was not carried out because of the relatively short time period between surveys and the relatively small number of field parcels (44) in which land cover change was recorded. Land cover transitions represented agricultural rotations, land cover conversion of seminatural grassland by agricultural intensification and land use change of seminatural grassland to conifer plantation. Annual rates of land cover change were calculated using a mean baseline date of 1990 (a monitoring period of three years). The types showing the greatest decrease in area were, species-rich grassland (5.7%) and fen meadow (3.3%).

Fig. 3. The mean number of species in 4m² quadrats against the mean number of nitrogen indicator species (Ellenberg 1988) in cluster analysis groups 1 - 15. Groups were named to facilitate comparison with grassland land cover types.

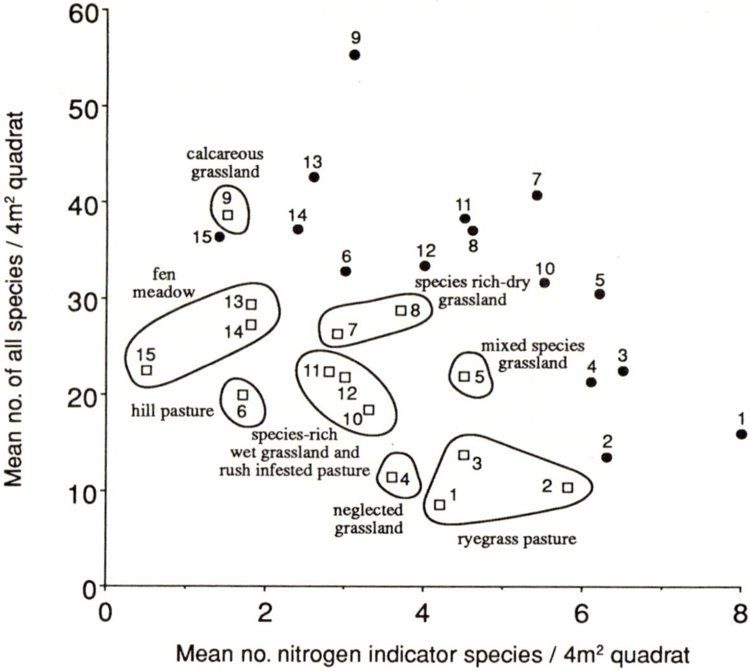

Discussion

The structured sampling programme within which grassland survey was carried out provides a framework for defining the composition of species-rich grasslands and estimating differences in the area of land occupied by different types of grassland. The sampling programme contributes to a formal definition of habitat size and quality and its spatial and temporal variation; major patch characteristics upon which species population survival depends (Selman 1993).

The main types of species-rich grasslands in the lowland landscapes of NI (species-rich dry and wet grasslands and fen meadow) are highly fragmented. Their patch structure differs from that of seminatural grasslands in upland landscapes (calcareous grassland and hill pasture), where the greater area and a larger parcel size gives a less fragmented grassland resource. This highlights the need to study the effects of fragmentation in relation to patch structure, landscape variation and vegetation community type.

There have been few plant ecological studies on the effects of fragmentation on grasslands. Emphasis has been largely on recording reductions in habitat area (Bunce *et al.* 1985) and recent decreases in the species diversity of grasslands (Barr *et al.* 1993). Island biogeographic theory (MacArthur and Wilson 1967) can be used to predict that small field parcels should contain less species than large parcels. There is little ecological data, however, to support this hypothesis or to explain the effects of fragmentation on grassland species diversity.

The results of our survey of Northern Irish grasslands suggest that species diversity does not decrease in small parcels and that the types of species assemblage do not change. The grasslands do not, therefore, seem to be influenced by fragmentation at the scale at which our studies have been carried out. The implication of this for the management of regional biodiversity, is that small parcels of species-rich grassland contribute significantly and that given appropriate agricultural management, can maintain themselves.

The strong relationship between species diversity and the number of nitrogen indicator species in the main types of grassland studied, emphasises the role that intensive agricultural management and grassland neglect play in determining their composition and diversity. Our results further suggest that under current agricultural management, conversion of species-rich grassland to more productive, less diverse agricultural grasslands, is likely to continue, probably until factors such as physical attributes of the landscape (e.g. parcel size, shape, soil drainage or stoniness) constrain management. The point at which this occurs is a function of landscape structure and farm management regime (Baudry *et al.* 1993).

A main conclusion of our work in Northern Ireland is that geographical factors are relatively unimportant in determining the diversity of species-rich grassland and that grassland management and its temporal and spatial variation are key factors. The effects of fragmentation on grasslands in farmed landscapes seems to be less obviously scale dependant or species-specific than with animal populations or vegetation types such as woodland. This is related to the size of woodland trees and the mobility of animals.

Patch structure and geographical factors such as isolation and connectivity may be more important in determining the species composition of more productive agricultural swards. The proximity of species sources such as field boundaries and neighbouring field parcels may be more influential in determining the diversity of grassland established on arable land, areas set aside from agriculture, or productive grassland which is managed less intensively.

References

Barr, C.J.; Bunce, R.G.H.; Clarke, R.T.; Fuller, R.M.; Furse, M.T.; Gillespie, M.K.; Groom, G.B.; Hallam, C.J.; Hornung, M.; Howard, D.C. & Ness, M.J. (1993) *Countryside Survey 1990: Main Report*, Department of Environment, Eastcote.

Baudry, J.; Thenail, C.; Le Coeur, D.; Burel, F. and Alaret, D. (1993) Landscape ecology and grassland conservation (Eds. R.J. Hagger and S. Peel) *Grassland*

Management and Nature Conservation. British Grassland Society Occasional Symposium 28: 157-166.

Bunce, R.G.H.; Crawley, R.V.; Gibson, R.A. and Pilling, R. (1985) *Composition of Enclosed Grasslands*, Yorkshire Dales National Park Committee, Bainbridge.

Cooper, A. (1986) *The Northern Ireland Land Classification*, University of Ulster, Coleraine.

Cooper, A & McCann, T. (1994) *The Botanical Composition of Grassland Land Cover Types in Northern Ireland*, Report to Environment Service, Department of the Environment for Northern Ireland, University of Ulster, Coleraine.

Ellenberg, H. (1988) *Vegetation Ecology of Central Europe,* Cambridge University Press, Cambridge.

Hill, M.O. (1979) *TWINSPAN,* Cornell University, New York.

MacArthur, R.H. and Wilson, E.O. (1967) *The Theory of Island Biogeography*, Princeton, New Jersey, Princeton University Press.

Murray, R., McCann, T. & Cooper, A. (1992) *A Land Classification and Landscape Ecological Study of Northern Ireland*, Report to Environmental Service, Department of the Environment for Northern Ireland, University of Ulster, Coleraine.

Selman, P. (1993) Landscape ecology and countryside planning: vision, theory and practice. *Journal of Rural Studies* 9, 1-21

SPSS (1990) *SPSS Reference Guide*, SPSS Inc., Chicago.

Wishart, D. (1978) *CLUSTAN User Manual,* 3rd edn., Inter-University/Research Council Series, Report No. 47, Edinburgh University, Edinburgh.

Fragmention patterns of ancient woodland in England

K.J.Kirby and R.C.Thomas

English Nature, Northminster House, Peterborough, PE1 1UA, U.K.

Abstract

Ancient woodland has long been a highly fragmented habitat compared to the original forests that covered England. This fragmentation has been increased by clearance and replanting over the last 50 years, although in places some sites are now less isolated than they were because of recent woodland development. Regional variation in the degree of fragmentation is very high. In samples from the High Weald there were 7 ancient woods per ten kilometre square over 50 ha, whereas on the Lancashire Plain samples there were only 3 in total per ten kilometre square and none were over 20 ha. The impact of fragmentation on ancient woodland species is affected by this regional variation and by the scale at which fragmentation is considered.

Introduction

Woodland once covered most of England, but had, in the counties covered by the Domesday surveys, already been reduced to about 15% by the eleventh century (Rackham, 1980). Clearance continued and total woodland cover was only about 5% by 1900. Much of this was not a remnant of the original natural forest, but new planting or natural regeneration that had occurred on open land during the previous centuries. During the twentieth century loss of woods that had survived from the middle ages and the development of new forests have continued. There has been a net increase to about 7% cover by 1990 (Forestry Commision, 1983) for all woodland types, but only about 2.6 % remains under ancient woodland, that is on sites that have been continuously wooded since at least 1600 AD (Spencer & Kirby 1992).

Results from the Ancient Woodland Inventory (Kirby *et al.* 1984) are used to describe fragmentation patterns for different parts of the country in terms of the size of wood, inter-wood distances and how patch size has changed since the beginning of the century. The scale at which fragmentation is considered has important implications for its effects on woodland species.

Methods

Ancient woods (defined as sites that have been continuously wooded since 1600 AD) have been identified for each county from old maps and other readily available historical sources. Their current state has been classified from aerial photographs and some field survey as either plantation or semi-natural (composed mainly of trees and shrubs native to the site that have not been planted) (Kirby *et al.*, 1984; Spencer & Kirby, 1992). Changes to the sites over the last 50 years approximately have been assessed by comparing their appearance on the first series 1:25,000 maps (mostly revised c 1930-40) with their state and size in the mid 1980s which is when most of the inventories were completed.

Fragmentation of ancient woods has been examined by comparing the size distributions of woods affected by clearance over the last 50 years across the whole country. Five of English Nature's "natural areas" (English Nature, 1993) were then chosen and three ten kilometre squares selected at random from each to look at fragmentation patterns in different types of wooded landscapes. In each ten kilometre square the cover and size distribution for all ancient woodland was calculated; then three ancient woods, three recent woods and three other points were selected at random. For each of these nine positions per ten kilometre square the distance to the nearest ancient wood and to a wood of any sort were measured.

There is much concern about the effects of roads on habitat fragmentation. The size distributions of woods affected by clearance associated with the development of the M25 orbital motorway round London before and after construction were compared.

Results

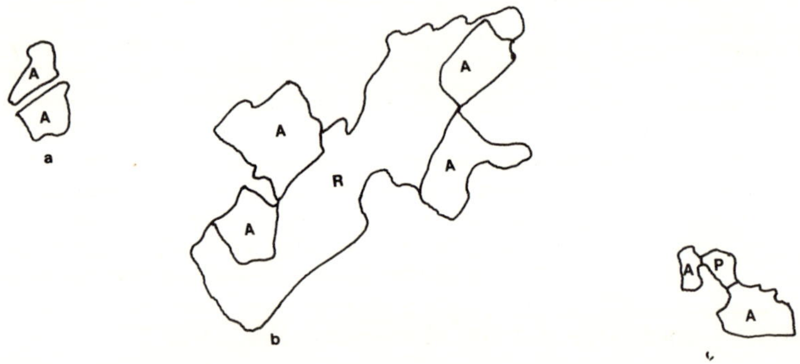

Figure 1. (a) Ancient woodland fragmentation at the site scale caused by a recent road scheme; (b) reduced isolation following new woodland development; and (c) increased isolation of semi-natural stands within ancient woods following replanting. A = ancient semi-natural, P = ancient replanted, R = recent woodland. Scale 1:50,000.

Many sites that were continuous in the early part of this century have become fragmented because of for example road schemes (Figure 1a), agricultural clearance and urban development. Relative isolation of sites has also increased because over 700 woods have been lost altogether, mainly to farming, and a further c4000 (19% of the total number of sites on the inventory) have been reduced in size (Table 1). The area of total clearance of ancient woodland in England since c1930 is about 7%. Some sites have however now been linked by new woodland development, in some cases through natural regeneration, in others, as in Figure 1b, through planting.

Fragmentation of semi-natural stands within ancient woods is even more pronounced with many smaller patches and fewer sites where it still survives (Table 2). Whereas in the early part of the century more than 1200 sites contained at least 50 ha of ancient semi-natural woodland, less than half that number still do.

Table 1. The size distribution of all ancient woods on the inventory that have been partly or wholly cleared (c1930 - c1980)

Size category (ha)	No of woods						
	0	1-5	6-10	11-20	21-50	51-100	101+
Area c1930	-	1460	1133	1039	823	315	202
Area c1980	759	1746	787	677	587	249	167

Table 2. The size distribution of all ancient semi-natural stands on the inventory compared to ancient wood size (c1980)

Size category (ha)	No of woods/stands						
	0	1-5	6-10	11-20	21-50	51-100	101+
Wood area	-	9992	4725	3339	2561	822	425
Semi-natural area	3765	10058	2886	2618	1039	368	131

Regional differences in fragmentation patterns for ancient woods

The survival of ancient woodland across the country is very patchy reflecting differences in the effects of soils, climate, history and land-use (eg. Marren, 1992; Rackham 1980). Attempts to assess the overall impact of woodland fragmentation on species must allow for the extreme variations in mean wood size, cover and number of woods per ten kilometre square and inter-wood distances (Tables 3, 4) between, in these examples, the High Weald (15% cover and nearly 100 sites per ten kilometre square, including 7 over 50 ha) and the Lancashire Plain (0.2% cover and only 3 small sites per ten kilometre square).

Mean minimum distances between such ancient woods as survive in the Lancashire Plain or the Border Uplands are surprisingly similar to those for more densely wooded regions because ancient woods tend to survive in groups. The difference is that in the Weald the ancient woodland is extensive and widespread so distances from recent woods or random points to ancient woods are also small, whereas these distances are much higher for the other regions.

Table 3. Percentage cover and size distribution for ancient woods in five natural areas (English Nature 1993), based on data from three randomly selected ten kilometre squares per natural area.

			No of woods in each size class (ha) per 10km sq					
Natural area	Ten km squares assessed	% cover	1-5	6-10	11-20	21-50	51-100	101+
High Weald	TQ93, TQ62, TQ52	15	41	14	12	11	5	2
East Anglian Plain	TF92, TL75, TM15	0.9	7	3	2	0	0	0
Exmoor	SS64, SS73, SS93	3.1	7	6	4	3	1	1
Lancashire Plain	SD33, SD41, SD45	0.2	1	1	1	0	0	0
Border Uplands	NY68, NY99, NU00	0.4	3	1	1	2	0	0

Table 4. Minimum distances between one wood or a randomly selected point and the nearest wood.
(Based on three ancient, three recent and three points in each of the ten kilometre squares used in Table 3).

	Minimum distances (kilometres) beween:					
	1	2	3	4	5	6
Natural area:						
High Weald	0.2	0.1	0.1	0.1	0.4	0.4
East Anglian Plain	0.6	0.4	2.8	1.2	1.9	1.2
Exmoor	0.2	0.1	1.2	0.4	1.3	0.5
Lancashire Plain	0.5	0.2	6.5	0.3	8.0	1.4
Border Uplands	0.6	0.1	2.6	0.2	6.1	0.3

1 = ancient wood to ancient wood, 2 = ancient wood to any wood; 3 = recent wood to ancient wood, 4 = recent wood to any wood; 5 = random point to ancient wood, 6 = random point to any wood.

The implications for the colonization of new woods in the Lancashire Plain are very different to those in the High Weald if the location of the new wood is effectively at random to existing woodland cover (ie equivalent to the point data in Table 4). In the Weald a species found in ancient woods might need to travel only 0.4 km from ancient source to new woodland, whereas in the Lancashire Plain it would be 8.0 km. However if the spcies were also in recent woods the colonization distance might be reduced to 1.4 km. A further consideration not obtainable from these data, but obvious from maps and on the ground, is that the minimum distance between woods in often very much less in one direction than another because particularly in the uplands woods are often arranged along linear features such as slopes or river valleys.

Fragmentation of woodland along the route of the M25 motorway

Clearance of ancient woods for roads or associated urban development along the route of the M25 motorway has caused a shift in their size distribution (Table 5). There are more than three times as many very small woods (less than 6 ha).

Table 5. Size distribution for ancient woods affected by the route of the M25 (c1930 - c1985).

Size category (ha)	No of woods in each category					
	0	1-5	6-10	11-20	21-50	51-100
Area c1930		8	8	9	9	2
Area c1985	1	29	14	6	7	2

Discussion

The data presented here present few surprises - the trends and specific examples have been presented elsewhere (eg. Peterken, 1981; Spencer & Kirby, 1992). However they start to quantify the high regional variation that exists in fragmentation patterns in ancient woods and illustrate the varying degree to which the increasing amount of recent woodland may be helping to reduce the isolation of surviving ancient woods. Our intention is to extend this analysis to other parts of the country.

Concern about the effects of fragmentation on the diversity of English woods must separate various aspects: the physical effects on the habitat, such as those described in this paper; the effects (if any) that splitting sites up or increasing the distance between them may have on species populations: and the effects of creating new edges which is important , but not considered further here. Similarly although we illustrate in Figure 1b an instance where new woodland has developed and linked two ancient woods, has this, in practice, helped to reduce the isolation of the species populations in the previously isolated sites; is a conifer plantation any less hostile for many woodland species than an arable field?

It may be helpful to consider three different points along the spectrum of potential isolation/fragmentation situations and their possible impact on different species: fragmentation of semi-natural stands within a site that remains completely wooded (albeit perhaps now mainly coniferous plantations); fragmentation of sites or other places where the distances between patches are quite small (tens or a few hundreds of metres) (Figure 1a, or the situation in the High Weald Table 4); and landscape-scale fragmentation where groups of woods (or even single woods) may be separated from others by several kilometres or more (eg. Lancashire Plain, Table 4).

Relatively little is known of the effects of fragmentation of semi-natural stands by plantations, although it probably is significant for some invertebrates, by analogy with what is known about the poor dispersal ability of the heath fritillary butterfly (*Mellicta athalia*) in coppice woods (Warren, 1987). Further discussion will therefore be limited to comparing site and landscape scales of fragmentation and site isolation.

It is helpful to consider three categories of species:
(a) those that would not move across the gap between patches whether it is wooded or not within the next fifty years;
(b) those that can readily move across the gap whether it is wooded or not;
(c) those that can and will move across the gap provided it is wooded.

Depending on the size of gap so species may fall into one or other of the categories. (The nature of the land-use within the gap is also important; some may be very much more hostile to movement than others for a particular species eg the difference between a road and a grass field of equal width. However such differences are a separate issue.)

Great-spotted woodpeckers (*Dendrocopos major*) appear to need a woodland area of about 10 ha for breeding (Hinsley *et al.*, 1994) but this can be in several discrete patches and the actual nest site may be in a fragment as small as 0.26 ha. Site-scale fragmentation may not affect them (unless an impassible barrier is created) provided the total area of woodland available remains above the minimum threshold. Equally creating links between two close woods, without substantially altering the total habitat area, might do little to improve their populations.

By contrast for relatively immobile plants and invertebrates site scale fragmentation or linkage of individual sites can effect their distribution, as they do spread into contiguous woodland (e.g. Webster & Kirby 1988; Peterken & Game, 1984). For these species changes in the relative isolation of sites at the landscape scale (eg changing the minimum distance between woods from (say) five to one kilometre) are largely irrelevant if only natural colonization processes are considered. The species are unlikely to move across the gap in either case.

This and other recent work for English Nature on habitat fragmentation and related issues (eg. English Nature 1994; Dawson, 1994, also this volume; Spellerberg & Gaywood, 1994; Hill *et al.*, 1993; Hinsley *et al.*, 1994) are leading us towards the following conclusions on the best ways of developing new woodland patterns through use of the forestry grants and other countryside management schemes.
(a) Where there is already a high woodland cover and intersite distances are relatively short (eg. the High Weald, Tables 3, 4) developing more is a relatively low conservation priority. The precise location of any new woodland in relation to other sites is also not that critical because

colonization is likely to be relatively fast for most species everywhere. Indeed if too many links are established between the existing woods there may be a risk that other habitats (eg. grassland patches) will become more fragmented. This in fact happened on many parts of the downlands of southern England in the 1950s following the spread of scrub and woodland after the decline of rabbits..

(b) In areas where woodland cover is low (the Lancashire Plain in Table 4) the emphasis should be on adding to existing sites and linking clusters to increase the potential population size of woodland species and to allow easy spread into the new woodland of poor colonists. New sites in the void between sites will do little to reduce overall inter-site distances and hence will not bring in specialist species not already present.

(c) If the aim is to fill in gaps in woodland patterns this may be done most effectively through a scatter of small sites rather than clumping it all together, the same conclusion reached for birds by Bibby *et al.* (1985) looking at the effects of broadleaved clumps in conifer plantations. It also reflects the value placed on hedgerows and other linear features simply as habitat in the countryside (not as corridors) which is out of proportion to their total area.

(d) If large new isolated woods are to be created, for example as part of major forestry commercial development or as for example in the New National Forest or Community Forests then these are not likely to be colonized naturally by specialist core species (or only very slowly) and it may be necessary to consider introductions.

The above are generalisations and do not take account of some of the other serious effects that may be associated with fragmentation, particularly increased pollution at the edges, whether from farmland or roads. Nevertheless we believe these to be a useful starting point for dealing with the variability in existing woodland landscape patterns and how we may improve them for nature conservation in future.

References

Bibby, C.J.; Phillips, B.N. & Seddon, A.J.E. (1985) Birds of restocked conifer plantations in Wales. *Journal of Applied Ecology* **22**: 619-635.

Dawson, D. (1994) *Are habitat corridors conduits for animals and plants in a fragmented landscape*? English Nature (Research Report 94), Peterborough.

English Nature (1993) *Strategy for the 1990s - natural areas,* English Nature, Peterborough.

English Nature (1994) *Roads and nature conservation: guidance on impacts mitigation and enhancement*, English Nature, Peterborough.

Forestry Commission (1983) *England - census of woodlands and trees*, Forestry Commission, Edinburgh.

Hill, M.O.; Carey, P.D.; Eversham, B.C.; Arnold, H.R.; Preston, C.D.; Telfer, M.G.; Brown, N.J.; Veitch, N.; Welch, R.C.; Elmes, G.W. & Buse, A. (1993). *The role of corridors, stepping stones and islands for species conservation in a changing climate*, English Nature (Research Reports 75) Peterborough.

Hinsley, S.A.; Bellamy, P.E.; Newton I. & Sparks, T.H. (1994) *Factors affecting the presence of individual breeding bird species in woodland fragments*, English Nature (Research Report 99), Peterborough.

Kirby, K.J.; Peterken, G.F.; Spencer, J.W. & Walker, G.J. (1984) *Inventories of ancient and semi-natural woodland*, Nature Conservancy Council (Focus on nature conservation 6) Peterborough.

Marren, P. (1992) *The wildwoods.* David & Charles, Newton Abbott.

Peterken, G.F. (1981) *Woodland conservation and management*, Chapman and Hall, London.

Peterken, G.F. & Game, M. (1984) Historical factors affecting the number and distribution of vascular plant species in the woodlands of central Lincolnshire. *Journal of Ecology* **72**: 155-182.

Rackham, O. (1980) *Ancient woodland*, Edward Arnold, London.

Spellerberg, I. F. & Gaywood, M.J. (1993) *Linear features: linear habitats and wildlife corridors,* English Nature (Research Reports 60) Peterborough.

Spencer, J.W. & Kirby, K.J. (1992) An ancient woodland inventory for England and Wales. *Biological Conservation* **62:** 77-93.

Warren, M.S. (1987) The ecology and conservation of the heath fritllary butterfly, Mellicta athalia. III Population dynamics and the effects of habitat management. *Journal of Applied Ecology* **24:** 499-513.

Webster, S.D. & Kirby, K.J. (1988) A comparison of the structure and composition of an ancient and an adjacent recent wood in Essex. *The London Naturalist* **67**: 33-45.

FRAGMENTATION IN AGRICULTURAL LANDSCAPES

Self-Organised Landscapes

Their Dynamics and the Implications for Management and Design

Simon Bell

Forestry Authority, Edinburgh, EH12 7AT, UK.

Abstract

This paper explores the evolution of landscape patterns from before exploitation by humans up to the time of organised planning. It argues that they were self-organised and as such well fitted to their surroundings, the climate and the processes at work. By including this understanding in a process for the planning and design of future agricultural landscapes the author argues that better ecological and visual results will be gained. A refinement of the design process is presented which uses a range of analytical methods to uncover the underlying, self-organised patterns as a basis for design.

Introduction

When we look at a landscape such as that of Great Britain where man has managed and altered it for thousands of years, it is easy to forget that it was once covered in completely natural vegetation, mostly different species of woodland (Tansley, 1949; Peterken, 1982; Rackham, 1986; Hoskins 1955). We see all around us the evidence of human control, planning and design. We are used to a pattern of intervention and order in the landscape which we may find reassuring and comforting (Crowe and Mitchell, 1988).

Over the last thirty or so years the landscape of the countryside has, in many places, changed dramatically (Blunden and Curry; 1988). Many of its features and patterns, often of ancient origin have disappeared to be replaced by visually bland and ecologically impoverished scenes. Many of us are now trying to restore some of the landscape and conservation values that have been lost, though not necessarily recreating exactly the previous landscapes. Various schemes funded by government grants: the Woodland Grant Scheme, Countryside Stewardship, Farm Woodland Premium Scheme, for example, enable trees and hedges to be planted or ponds to be created. Landowners in Environmentally Sensitive Areas can be paid to carry out traditional forms of management to protect flora and fauna. Areas of wetland, water, woodland and grassland have become nature reserves and are managed for their wildlife or nature conservation values.

While all this activity, worthy as it is, goes on, it remains in many ways an *ad hoc* process. Sites may be looked after because they are strategic but much happens piecemeal, at the whim of the landowner and without the framework of a broad view in which to fit the detail. If we look closely at the evolution of any landscape we find that conscious planning or design are mostly relative newcomers to the scene. The landscapes of many areas of Britain have gradually evolved over the years and consequently have achieved a goodness of fit and unity which is difficult to achieve by design. We might term these 'self-organised' landscapes. Even when human use and alteration is taken into account they can be called self-organised because they exhibit a constancy and continuity of pattern over wide distances in the absence of a single guiding hand (Bell, 1994 in press).

This paper explores this phenomenon and offers a revised design process aimed at helping to halt and reverse the fragmentation of agricultural landscapes. This process works with the dynamics of nature and develops a sensitivity for the land which has become all but lost.

FRAGMENTATION IN AGRICULTURAL LANDSCAPES

The Patterns of Nature

Recent research into the dynamics of natural systems is coming to understand something of the way incredibly complex, highly ordered patterns seem to appear from apparently random circumstances, this is known as the chaos theory (Gleick, 1988). Among the range of phenomena to which this has been applied are weather systems, fluid dynamics, stockmarket fluctuations and plant growth. The underlying concept emerging from this work is that while it may often seem chaotic and unstructured the world and the processes at work in it exhibit a deep underlying order, often repeated at a range of scales from very large to very small. For example, we can see in the endlessly repeated patterns of ripplemarks in sand or cracks in dried mud an almost perfectly repeated pattern, incredibly detailed, highly ordered yet not consciously designed. It is the interaction of the medium, sand or mud, with the processes, wave movement or the drying heat of the sun and the dynamics behind these processes which cause this. The aesthetics of such patterns can often be extremely beautiful (Gleick, 1988).

When Britain was re-colonised by vegetation after the last ice age there developed a complex pattern ranging from the large-scale distribution of woodland types down to the local, detailed structure of trees, shrubs and ground cover plants. The coarser grained patterns would have reflected latitude and climate: the wetter, cooler north-west grading to the drier, warmer south east, for example. Geology, landform and soils would have added more pattern at a finer scale, for example chalk, or limestone, high mountains or low-lying valleys. This pattern did not stabilise into climax vegetation in quite the way some have supposed (Tansley, 1949) but developed a dynamic equilibrium where, with climatic fluctuations the proportion of different species in different areas altered, ebbed and flowed to a greater or lesser extent over time.

At the finer grain, within a given woodland area, there would have occurred more local variation due to microsite and microclimate, for example wetter flushes, drier knolls, nutrient receiving sites or outcropping rock which led to the development of patches of different plant communities within the general matrix of woodland. Such patterns can be picked out using the key from the National Vegetation Classification (NVC) (Rodwell, 1991). Surveys prepared on this basis are already being used as a starting point for the layout of new native woodlands (Rodwell and Patterson, 1994).

The Pattern of Cultivation

When the early farmers arrived they started clearing the woodlands for agriculture, probably choosing their settlement according to a variety of factors: accessibility from rivers or trackways, protection from floods or attack, ease of clearance using primitive tools and traditions brought with them for the layout of farms and settlements. It is conceivable that families or family groups settled in an area and carved their fields out of the wildwood much as the early European settlers did in eastern USA in the 1600s and 1700s. They would have taken common site factors and physical limitations into account, so that while there was no higher authority directing their colonising efforts there would have been a large degree of similarity in the resulting patterns (Bell, 1994). Soil, shelter, insolation and drainage would have been important limiting factors for the carrying capacity of the landscape and the types and yields of crops or animals that could have been grown. Equally, the people would have used the materials close to hand for buildings and so on. Thus these small family groups, widely scattered at low population density, developed the landscape so that over a relatively short period of time, some 2-3,000 years (Rackham, 1986), significant areas of consistent patterns of cleared land emerged out of the formerly wooded landscape.

Surprisingly large areas of these patterns actually persist today, many in quite good condition, especially in the south and west where pastoral agriculture remains important. These are the 'ancient' landscapes of England and Wales (Rackham, 1986). They are characterised by small, irregularly shaped fields, winding lanes, dense hedgerows composed of many species, ancient semi-natural woodlands and clustered settlements (Forestry Commission, 1992). By the definition explained above, they are also self-organised landscapes.

A further aspect of many such landscapes is that they have not remained entirely static, particularly at the margins where colonisation was later and where the population was most vulnerable to famine, climatic fluctuations and economic hardship. These, often the upland fringe areas, have seen agriculture advance and retreat, often several times over the centuries (Hoskins, 1995; Bell, 1994). There may be old field boundaries present to this day on what is now heathland or moor showing where farming was once possible. In other places some of these old fields have been reseeded as late as the early 1980s after being heathland for centuries.

Woodland often occurs between enclosed fields and moorland, especially where the moor has always been common grazing and unenclosed. This might have been managed as wood pasture or coppice. Over time it may have moved up or down slope, perhaps when grazing pressure was light allowing for regeneration to be successful, or when extra fields were carved out along the lower margins (Bell, 1994). Where this fluctuation took place would depend in part on the local landform and soils and was not a random movement. These can also be described as self-organised, dynamic patterns. Such circumstances also show that preserving landscape patterns is not necessarily an appropriate course of action, because changes in various ways are normal and to be accepted.

Another feature of many natural and man-made patterns is their fractal character. This represents another aspect of chaos theory: the self-similarity of patterns at a range of scales which leads to a highly interlocked result with a lot of edge (Mandelbrot, 1977; Gleick, 1988). Such characteristics may result from the underlying landform and its own hierarchy of structure, for example dendritic drainage patterns. The transition zones from one pattern to another often display fractal geometry most obviously, for example a natural upper tree line, woodland spreading on to a bog or vegetation in intertidal mud flats. Once again, these patterns are not static, they are dynamic, ebbing and flowing over varying periods of time.

How Can Natural, Self-Organised Patterns be Maintained?

If many landscapes have shown characteristics of self-organisation in the past by farmers working the land and as a result have fitted extremely well into their time and place, then why should we be so concerned about planning and design? Is it not possible to allow individual farmers' actions to continue to determine the landscape? There are four reasons why this is not possible and why we should intervene.

1. The 'Cartesian world view' has led, since the Age of Enlightenment, to a belief that man can control nature and impose his will over it (Harrison, 1992). This has led to abstract, mainly geometric patterns being superimposed as the result of a single decision instead of the aggregations of site sensitive, multiple decisions characteristic of ancient landscapes. This is best illustrated in the planned landscapes of parts of Britain where the geometric field layout is most obvious. It is interesting to note, however, that the distinction over time between the two is not clear cut: there were some extremely early examples of planned landscapes, some of the most well known of which are the Dartmoor Reaves (Fleming, 1988), where a single decision must have been made in some way.

2. Centralised agricultural and land use planning delivered via grants and subsidies has tended to override the decisions of individual farmers whose choices of action are inevitably more concerned with maintaining their farm incomes than with the traditional view of good husbandry. This is leading to a degree of homogenisation of the landscape (Bell, 1994).

3. Industrialisation and the spread of agricultural technology has reduced farmers' reliance on the subtleties of the land and soil. Fertilisers, drainage and genetically improved crop varieties have enabled more crops to be grown on a wider range of land. Mechanisation has led to larger fields and a simpler landscape at a coarser scale (Blunden and Curry, 1988).

4. The increased interest in the countryside as a place to visit by mainly urban people who are not active participants and the decrease of numbers of people who live and work there has changed the balance of public perception. This has affected the degree to which certain management practices or aspects of landscape change are acceptable (Crowe and Mitchell, 1988).

It is not practical, in the interests of landscape or nature conservation, to revert to old-style labour intensive agriculture. Farm incomes and the loss of rural labour will not allow it, so management in terms of fewer numbers of workers per hectare and mechanisation will continue to be the rule. This means that if certain patterns are shown to be ideal for ecology and aesthetic quality they will have to be planned and designed so as to be able to work within this context. This in turn means that the design process as we currently understand it needs some adjustment or refinement while a wider vision is needed to escape from the strait-jacket of site protection as the main tactic.

The task is potentially large as there are many landscapes which are highly fragmented. Woodlands have been isolated as hedgerow connections have been broken. The same has occurred for patches of grassland, heath, ponds and wetlands. This has also had a profound effect on the landscape character, reducing unity and diversity. Since it has been perfectly possible for a wide range of plants and animals to survive, adapt and thrive in the finely woven fabric of the ancient agricultural landscapes we can analyse their components and find out what is characteristic so that by design it can be replaced.

An analysis using landscape ecological principles can help us to understand how such structures work at the broad scale. It has been proposed that we should concentrate on the structure, components and dynamics of the landscape on the assumption that if this is resolved then all the appropriate species which inhabit it and are available for recolonisation will be catered for (Diaz and Apostol, 1992). The balance between the integrity of the pattern, the overall level of structural diversity, the ratios of edge length to area (fractal dimension) and the dynamics of landscape evolution determine how and at what level the landscape is ecologically sustainable. If this is put right then the detailed site level and operational activities will be simpler to effect. A hierarchical approach is worth adopting, starting with the broadest, progressively working down to the finest scale.

Landscape Ecological Analysis

The following section describes briefly how the landscape can be analysed to help understand how it functions ecologically. It is based on a 4 step approach:

1. Describe the landscape structure. This entails dividing the landscape into the components of matrix, patches and corridors (Forman and Godron, 1984; Diaz and Apostol, 1992). The matrix is that part of the landscape which has most connectivity, the greatest relative area and contributes most to the landscape dynamics. In many agricultural landscapes this might usually be extensive areas of open fields. These may be less matrix in some areas where the green fields are divided by strong hedgerows and other features so that

it may be more or less absent at the broadest scale. At finer scales a single field may be a matrix. In a woodland the continuous canopy is the matrix and is usually inhabited by plants and animals who require 'interior' habit (low light levels, lower temperature fluctuation, high relative humidity). Other matrix examples could be heather moorland or chalk downland where open habitat species thrive.

Patches are areas which are relatively homogeneous in character and contrast with their surroundings (the matrix, other patches). Small woodlands, ponds, small areas of heath in arable land are examples which may often be where most ecological and visual diversity is greatest. Many patches are man-made, such as surfaced areas, settlement, quarries. These add to the fragmentation of the landscape. Some areas may be virtually all patches if there is no component large or continuous enough to be called matrix. This may increase overall diversity but there may be problems if there is no connection between any similar patches. Edge species will be catered for primarily.

Corridors are linear features that connect similar patches through a dissimilar matrix or aggregation of patches. Common examples in agricultural landscape are hedgerows, grassy lanes or streams. These are usually important for linking otherwise isolated elements such as woodlands or unimproved grassland. Other corridors act as barriers to many kinds of wildlife, for example motorways or railways.

The precise relationship between each of the above components may depend on the scale and the historical development of the landscape. In an ancient landscape the elements which are described as patches and corridors, the hedgerows and wooded areas, are possibly the last remnants of the original forest matrix. The fields could well have started as small clearings and gradually increased in number and size over the centuries. In other areas open country was enclosed by hedges and plantation woodlands created. Thus in the first example fragmentation of the pattern removes the last link to the past continuity while in the second the return to openness is historical except that the vegetation of the matrix is probably quite different.

2. Describe the flows in the landscape. Flows are those agents, living or non-living which move across (on land, in the air, in water or in the soil) the landscape. Animals, bird, fish, water and humans are typical examples. At the large scale to flows of interest are those which move around the whole area, while at the smaller scale flows might be much more limited. The range of birds or deer can be compared with that of slugs or worms, for example.

3. Each of the flows relates to or is influenced by different components of the structure. Some may need a range of elements, perhaps for nesting, hunting over or roosting in. Others will be confined to one element, perhaps open heathland or interior shady woodland. It is useful to compile a matrix or spreadsheet to show the relationships involved. From this it is an easy step to see what elements of the structure are absent or under-represented, what flows are presented with optimum conditions or are under pressure.

4. The next step is to explore the dynamics in the landscape. This can be sub-divided into disturbance factors and natural succession stages. Disturbance factors can be natural - such as wind, fire, insect attack or flood, or human, such as cultivation, woodland management, quarrying or development. Succession is the process by which a disturbed area reverts to some kind of 'climax', that is the most highly developed seral state climate, soil and disturbance will allow. It might be mature native woodland or heathland, for example. In the context of a changing landscape the past dynamic processes can be compared with the current or potential future ones to assess what impact they have. Previous farming practices may have encouraged certain vegetation while current or future ones might cause different species to grow. Natural land may have reverted to woodland if

farming was abandoned whereas gravel quarrying could leave water behind and start a completely different successional process.

At the end of this step a picture should have been built up as to the current state of the landscape and the trends for change in terms of its ecological functioning. This can be presented as a series of map overlays, spreadsheets and descriptions. From it a model can be developed which tells us how to manipulate the landscape in order to enhance or maintain its functioning over time depending on various levels of input. Modified management practices, agricultural systems or abandonment to 'nature' might be possible scenarios. In addition a model of the optimum landscape structure can be developed including restoration of missing components, mimicking natural disturbances and creation of new patterns. After this the ecological model has to be included in the comprehensive design process.

A New Approach to Design

The method of design proposed has aims to use the understanding of the dynamics of self-organised landscapes, the landscape ecological analysis and the creative landscape design process. The objective is to ensure that agricultural landscapes are:

- able to be managed by farmers to produce food economically;
- ecologically sustainable;
- reflect the appropriate landscape character.

The steps of the design process are as follows:

1. Identify the ecological/landscape character zone for the landscape.

2. Prepare an analysis of the actual or potential National Vegetation Classification for the area.

3. Using a combination of landform, soil patterns and historical landform information, determine the characteristics of the landscape patterns and the degree of self-organisation.

4. Prepare the landscape ecological analysis.

5. Develop a concept for the development of the landscape structure starting with the coarsest scale and the largest size of appropriate habitat components.

6. Work down from the coarse scale to progressively finer scales building in more subtlety as each scale is reached.

Step 1. Identify the ecological/landscape character zone.

This may be achieved by reference to the English Nature map of natural areas (English Nature, 1993), a landscape assessment compiled according to currently recommended practice (Countryside Commission, 1993; Price, 1993) or be compiled specially. The key requirement is to set the context, the general kind of landscape, its importance, representativeness and the main ecological classification present based on soil, geology, vegetation, landuse and landform types present.

Step 2. Prepare a National Vegetation Classification analysis.

This will show what is present at the moment and what the categories might be if allowed to develop. Such an analysis should show where and how much of each and therefore what would be most likely to be naturally the most ecologically sustainable landscape structure in terms of management and natural levels of diversity.

Step 3. The landscape pattern and dynamics of self-organisation.

Using the above information, old maps showing field boundaries and woodland (perhaps early series 2½ inch Ordnance Survey maps) and photographs from typical viewpoints, analyse the landscape in terms of its visual character, that is the types of shapes of features, their relationship to soils, landform and climate. This could show, for example, where and why woodlands were left uncleared in the past, the size and shape of fields and how these relate to the underlying factors or the changes in landscape progressing up a hillside. This analysis should then show where it would be most 'logical' to extend woodland, to recreate wetland, to revert areas to acid grassland within the overall context of the pattern rather than in ad hoc fashion as land becomes available. In the case of planned landscapes or areas where most of the features have disappeared detecting the earlier, underlying pattern will be necessary. This approach does not argue for wholesale restoration but often for new landscapes created in the spirit of the inherent character of the landscape. It is this kind of area where the use of the 'self-organised' approach may be most fruitful.

Step 4. Prepare a landscape ecological analysis.

The method for compiling this has been described above. Once completed the ideal ecological model should be compared with the landscape pattern analysis from Step 3 to see how far the two can be merged together. The greater the fusion, the more subtle and sustainable the end result will be.

Step 5. The Ecological Landscape Design Concept.

This step emerges from step 4. The layout at the coarsest scale of the largest, most significant components should be planned. This should include the concept for their linkages, for example, river corridors, large ancient woodlands with stable interior habitat, large new native woodland areas, recreated water, heathlands or grassland. These should reflect the major NVC types and be of sufficient scale that each element could be managed through a full cycle of succession over a long period of time. The location of new features should be decided with reference to the landscape pattern showing how natural colonisation might be likely to proceed. These could include old former sites of semi-natural habitat recently cleared which could be restored and extension from existing areas in organised fashion including corridor linkages to new areas.

Step 6. Progressive refinement down the scale.

The final step is to refine the broad strategic plan described in step 5 by progressive addition of detail, for example, smaller areas of woodland connected by corridors into the main structure, the detail of woodland layout, edge design and so on. This is where the fractal structure comes in. Using the repetition of the types of shape at each successive scale a closely, intricately woven tapestry will be created, right down to the finest detail if so desired. The degree of detail actually incorporated will depend on how complex the result would be to manage or how much additional benefit it gave to ecological functioning and aesthetic quality. What is certain, however, is that whether the landscape is perceived from external views looking on to the pattern, or from inside, say walking through it, the pattern will read and be comprehensible. The detail will obviously fit into and be related to the broader structure. Animals that use small, intricate parts will find their habitats ideal as will those who range more widely.

CONCLUSION

The design approach described above endeavours to move away from any approaches which try to preserve evolving landscapes, rely on ad hoc occurrences somehow coming together to create a whole, or seek to impose a preconceived style on the landscape. The designer becomes a conduit through which the refined understanding of the landscape and its dynamics passes and is transformed into a blueprint for creativity and development at a level of ecological and aesthetic sophistication rarely seen. The landscape can remain a truly living entity in which human activity can proceed without inevitably coming into conflict with the natural world. Hence the approach allows for true sustainability to be achieved.

REFERENCES

Bell, S. (1994) *New Woodlands in the landscape in The Ecology of Woodland Creation* (ed R Ferris Khan). John Wiley and Sons Ltd, London (in press).

Blunden, J. & Curry, N. (1988) *A Future for our Countryside.* Blackwell, Oxford.

Countryside Commission (1993) *Landscape Assessment: New Guidance Cpp 423.* Countryside Commission, Cheltenham.

Crowe, Dame S. & Mitchell, M. (1988) *The Pattern of Landscape.* Packard Publishing, Chichester.

Diaz, N. & Apostal, D. (1992) *Forest Landscape Analysis and Design.* US Forest Service, North West Pacific Region, Portland.

English Nature (1993) *Natural Areas: Consultation Document.* English Nature, Peterborough.

Fleming, A. (1988) *The Dartmoor Reaves*, Batsford, London.

Forestry Commission (1992) *Lowland Landscape Design Guidelines.* HMSO, London.

Forman, R. T. T. & Godron, M. (1986) *Landscape Ecology.* John Wiley and Sons, New York.

Gleick, J. (1988) *Chaos.* Heinemann, London.

Harrison, R. P. (1992) *Forests: the Shadow of Civilisation.* University of Chicago Press, Chicago.

Hoskins, W. G. (1955) *The Making of the English Landscape.* Hodder and Stoughton, London.

Mandelbrot, B. (1977) *The Fractal Geometry of Nature.* W H Freeman, New York.

Peterken, G. F. (1981) *Woodland Conservation and Management.* Chapman & Hall, London.

Price, G. (1993) *Landscape Assessment for Indicative Forestry Strategies*, Forestry Authority England. Cambridge.

Rackham, O. (1986) *The History of the Countryside.* Dent, London.

Rodwell, J. (1991) (Editor) *British Plant Communities.* Cambridge University Press, Cambridge.

Rodwell, J. & Patterson, G. (1994) *Creating New Native Woodlands.* Forestry Commission Bulletin 112, HMSO London.

Tansley, G. (1949) *Britain's Green Mantle.* Cambridge University Press, Cambridge.

Landscape design: preliminary estimates of the effects of landscape permutations on wildlife.

T.H. Sparks, S.A. Hinsley, J.O. Mountford, N. Veitch, P.E. Bellamy & D.S. de Nooijer

NERC Institute of Terrestrial Ecology, Monks Wood, Abbots Ripton, Huntingdon, Cambs PE17 2LS, UK

Abstract

Recent and planned agricultural de-intensification may result in a greater emphasis on managing the landscape for amenity and for conservation. There is an increasing need to evaluate the consequences of various management options. Computer modelling of species numbers and distributions combined with the facilities of Geographic Information Systems (GIS) provides such an opportunity.

In this paper we demonstrate a pilot study modelling the consequences of different landscape permutations on breeding birds and other wildlife. These reveal the vast potential for integrating ecological data from intensive surveys with computer modelling and GIS.

Introduction

Over recent years there have been a number of intensive studies of the ecology of separate landscape components. Such work has been carried out by a variety of organisations including research councils, government bodies, conservation groups and university departments. Funding for this work has been equally diverse. Progress in both computer modelling and in Geographic Information Systems (GIS) means that it is now practical to combine information from surveys of components to estimate the effects of manipulated whole landscapes on wildlife.

The British landscape has historically been shaped by man reacting to economic pressures. Agricultural de-intensification would allow some prospect of redesigning landscapes for both public amenity and wildlife. Computer modelling allows us to estimate the consequences of various permutations of landscape design on wildlife without trial-and-error experimentation and consequently it can be a relatively quick and inexpensive option. Such modelling can incorporate information from surveys and the considered judgement of workers in appropriate disciplines. This paper reports some preliminary results arising from work at ITE Monks Wood, and uses database information on three species groups (birds, butterflies and wetland flora) for illustration.

Landscape models

The landscape unit investigated is an hypothetical 2km x 2km square (400 hectares) in eastern England. Innumerable permutations of land use in such an area are possible but for reasons of brevity, we only discuss five possible scenarios. In summary these are:

I. A totally arable landscape divided, for historic reasons, into fields of varying shapes and sizes. Where hedgerows exist along field boundaries these tend to be heavily trimmed (to less than 2m in height). A field ditch system connects with two larger drains. Trees occur at a relatively low density along boundaries.

II. As I, except approximately 15% (c. 60 ha) of the area is dedicated to pastoral activity, and a small number of ponds are located within these pasture fields. Hedgerows adjoining pasture are less intensively managed (typically between 3m and 5m in height) and more hedgerow trees exist.

III. As II, except that the pasture exists in a single block.

IV. As II, except approximately 10% (c. 40 ha) of the area has been nominated as woodland.

V. As IV, except that the woodland exists in a single block.

These five scenarios are shown in Figure 1 and their main features are summarised in Table 1.

Table 1. Land use and field boundary statistics for the five example scenarios.

		I	II	III	IV	V
Areas(ha) of	Arable	400	338	339	294	294
	Pasture	0	62	61	62	62
	Woodland	0	0	0	44	44
Field boundaries(km) of						
	Arable	25.8	15.6	19.0	10.5	12.4
	Arable/Pasture	0	9.8	3.6	9.2	8.7
	Pasture	0	0.4	3.2	0.4	0.4
	Woodland edge	0	0	0	5.6	2.6

Figure 1 (opposite). The five scenarios modelled in this paper displaying the pattern of land use, field boundaries and water bodies. Unless otherwise indicated, land is dedicated to arable production. Top row - scenarios I and II, middle row - scenarios III and IV, bottom row - scenario V.

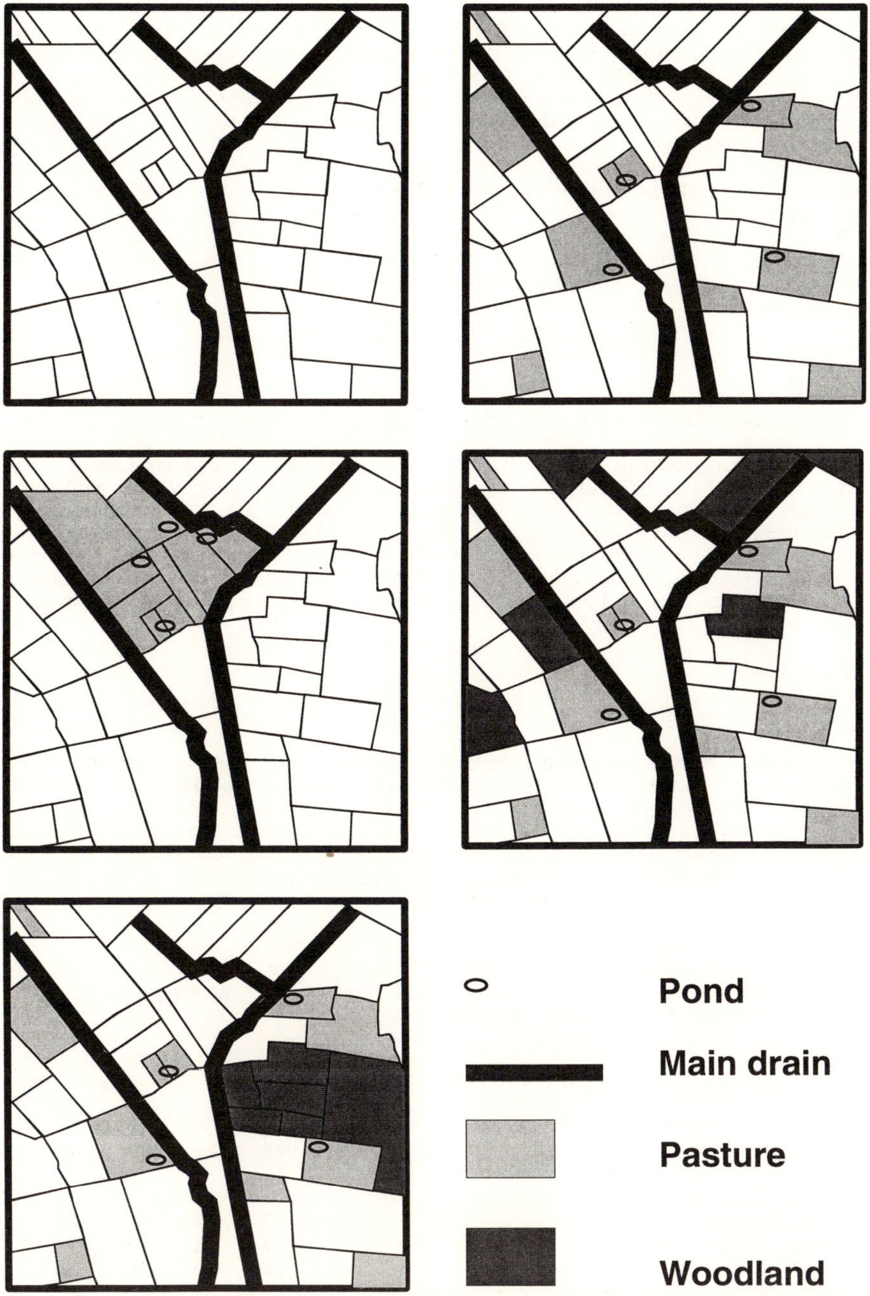
Pond
Main drain
Pasture
Woodland

The landscape models were held in a digital form within a GIS. A GIS is a suite of computer software used for the capture, storage, manipulation, display and analysis of spatial data. Within a GIS each landscape feature is represented by a series of spatial co-ordinates, a unique identification code and a series of attributes (e.g. field type, boundary type). The GIS was used to measure the length of the field boundaries and the area of each land use type for each scenario. The results of this were then used as an input to the modelling procedure.

In the case of the woodland birds, the estimates of numbers were obtained by a simple equation of the form

$$\hat{y}=\alpha+\sum \beta_i x_i+\sum \gamma w_j$$

where x_i and w_j represent the length of various field boundary types and the areas of individual woods respectively.

These models make a number of simplifying assumptions. For instance field boundaries are assumed to be fixed and not removable. This was felt to be reasonable in a comparison of landscapes differing in land use. The removal of such an assumption increases the number of possible permutations dramatically. It should also be noted that, in this preliminary assessment, no allowance has been made for buildings, gardens or roads.

Results

Woodland birds

Estimates of breeding bird densities of selected species are based on recent surveys of farmland and woodland in a year (1991) when population levels were generally low (Marchant & Musty, 1992). Because populations can fluctuate greatly from year to year more attention should be given to comparing the relative abundance between different scenarios than to the absolute values themselves. Results for eight species are presented in Table 2.

Table 2. Number of breeding pairs of eight bird species for each of the five example scenarios

	I	II	III	IV	V
Wren	6	16	13	61	59
Robin	3	8	10	45	45
Blackbird	16	21	22	43	40
Dunnock	19	21	21	38	34
Song Thrush	0	0	1	6	6
Blue Tit	0	3	2	38	37
Great Tit	3	3	4	38	34
Chaffinch	13	15	14	50	48

The most noticeable difference between the five scenarios is the dramatic increase in numbers of all species (except Song Thrush in which the increase was smaller, see below) following the addition of woodland to the landscape. The lack of cover for foraging and nesting and the limited availability of food in the arable landscape make it largely unsuitable for many birds typical of woodlands, well grown hedgerows and gardens. Only those able to forage in the open landscape (Blackbird and Chaffinch) or which can use the low cover of the field boundaries and ditches (Dunnock) occur, and then at relatively low densities (e.g. at c. 0.6 pairs km^{-1} of field boundary in scenario I). For the species used in this example, the increase was similar for the addition of woodland in both small units and as a single block. However, certain species typical of woodland edges, e.g. Yellowhammer, tend to be less common in the edges of large woods than of small ones, and hence may show a decrease in numbers as woodland size increases greater than that expected from the reduction in length of woodland edge alone.

This preliminary model does not demonstrate the importance of woodland to the more specialised woodland species, such as Treecreeper, Marsh Tit and Chiffchaff, but these data should become available in the future. Because these species occur at lower densities than those used in this example, their responses to increasing amounts of woodland in the landscape are likely to resemble that of the Song Thrush, which also occurs in relatively low numbers in our database woods, rather than that of the more numerous Blackbird. In contrast to the suggestion that edge species may be less numerous in a more heavily (and contiguously) wooded landscape, the specialists are likely to show the opposite trend of greater numbers in large woodland patches.

Butterflies

Eastern England has a smaller pool of butterfly species than the more wooded areas of southern England where Dover(1994) has recently summarised the species observed in arable farmland. With butterfly species it is more difficult to determine breeding than with, for example, birds. Presence alone may be insufficient to prove breeding, especially with the more mobile species. However, the detection of less mobile species, such as the Gatekeeper, are a good indication of breeding activity within the area. Table 3 lists those species that might be observed under the different scenarios.

The inclusion of woodland or pasture can substantially boost butterfly populations by perhaps half-a-dozen species. Metapopulation dynamics (Hanski & Thomas, 1994) may be critical in determining the persistence of individual species in a fragmented landscape. A different suite of species (and different relative abundances) would be expected in different regions of Britain, for example in southern England (Dover, 1994)

Table 2. Expected butterfly species within the five examined scenarios

	I	II	III	IV	V
Small White	✓	✓	✓	✓	✓
Green-veined White	r	r	✓	✓	✓
Large White	r	r	✓	✓	✓
Orange Tip	r	r	✓	✓	✓
Brimstone		r	r	✓	✓
Peacock	✓	✓	✓	✓	✓
Comma	r	r	r	✓	✓
Painted Lady	r	r	r	r	r
Red Admiral	✓	✓	✓	✓	✓
Small Tortoiseshell	✓	✓	✓	✓	✓
Ringlet	r	r	r	✓	✓
Gatekeeper	✓	✓	✓	✓	✓
Small Heath		✓	✓	✓	✓
Meadow Brown	✓	✓	✓	✓	✓
Wall Brown	✓	✓	✓	✓	✓
Small Copper		r	✓	✓	✓
Common Blue		r	r	✓	✓
Holly Blue		r	r	✓	✓
Small Skipper	r	r	✓	✓	✓
Essex Skipper	r	r	✓	✓	✓
Large Skipper	r	r	r	✓	✓
Speckled Wood				✓	✓
Green Hairstreak					r
Purple Hairstreak					r

✓: present; r: rare, present in low numbers. Painted Lady and Red Admiral are migrants and numbers can fluctuate greatly from year-to-year.

Wetland flora

Estimates of the wetland species numbers associated with different landscape permutations were derived from detailed studies and surveys of arable fenland, grazing marsh, ESA's and Set-aside. The numbers of aquatic macrophytes are directly linked to the depth and persistence of open water whilst the presence of truly wetland species in grassland is linked partly to the field water-table and partly to the presence of refugia for such species on ditch and pond banks (Mountford & Chapman, 1993).

In a purely arable landscape (Scenario I) only those species of aquatic macrophytes which can tolerate the elevated nutrient status linked with intensive arable systems will survive. A few macrophytes may persist in seasonally wet field ditches, where wet grassland species may also linger, although such plants are more widespread on the banks of permanent ditches and drains (Table 3).

Table 3. Number of plant species predicted to occur in each of the five example scenarios.

	I	II	III	IV	V
Aquatic macrophytes - emergent, floating and submerged in standing or slowly flowing water	18	42	50	42	42
Species of wet mesotrophic grassland and the banks of drainage channels or ponds	15	35	45	35	35

The other four scenarios contain additional wetland habitat in the form of a small number of ponds set within pasture. Such aquatic habitats may be partly protected from the impact of eutrophication, being isolated from the drainage network and contained within a lower fertiliser regime. The banks of ponds support a varied flora where species typical of a variety of water-regimes may exist at different levels up the bank.

Grazing can have a major impact on species richness by suppressing competitive species, enabling more light to reach shallow water bodies and creating bare ground opportunities for colonisation. However, much grassland in eastern England has been improved and no longer supports a diversity of wet grassland plants. These factors combine to make Scenarios II, IV & IV richer in wetland plants than Scenario I, though not in themselves distinguishable from one another in terms of species richness. The presence of large blocks of pasture (Scenario III) is frequently linked with less intensive farming regimes and, in particular, less rigorous attempts to lower the water-table.

Discussion

The concept adopted in producing this paper has included a number of simplifying assumptions, but this is not unusual in computer modelling. We stress that this work should be viewed as an examination of the potential of such an approach rather than as a definitive methodology. The advantage of using a GIS-based approach is that the analysis can be easily and rapidly extended to cover areas far greater than the 400ha used in this study. In addition, the results of the modelling itself may become attribute information held within the GIS to allow ecological considerations to be represented within the land-use planning process.

Each real situation is unique and should be treated on its own merits and each may require a modification of a more general approach. Modelling of permutations should take into account both practical and local restrictions. For instance, our scenario II (a fragmented arrangement of pasture fields) may be impractical for an intensive livestock farmer, whereas scenario III would be more practical. From a non-economic viewpoint, it is often the patchwork of a fragmented landscape that is more aesthetically pleasing, but such landscapes might not be able to sustain a viable population of a species. An examination of metapopulation dynamics may be necessary.

Despite its shortcomings, this work suggests that dramatic improvements to wildlife populations could be expected from relatively small changes to a basic landscape pattern. In our example scenarios, arable production has occupied at least 75% of the total area so maintenance of an economically viable countryside together with benefits to wildlife should be achievable.

Acknowledgements

The authors would like to thank Julie Gaunt for invaluable assistance and Tina Yates for useful discussions during the preparation of this paper.

References

Dover, J. (1994) Arable field margins: factors affecting butterfly distribution and abundance. N.D.Boatman (Ed). *BCPC Monograph no. 58, Field margins: integrating agriculture and conservation*, BCPC, Farnborough, 59-66.

Hanski, I. & Thomas, C.D. (1994) Metapopulation dynamics and conservation: a spatially explicit model applied to butterflies. *Biological Conservation* **68:** 167-180.

Marchant, J. & Musty, L. (1992) Common Birds Census: 1990-91 index report. *BTO News* **182:** 9-12.

Mountford, J.O. & Chapman, J.M. (1993) Water regime requirements of British wetland vegetation: using the moisture classifications of Ellenberg and Londo. *Journal of Environmental Management* **38:** 275-288.

Identifying landscape sensitivity through landscape character assessment with special reference to Scotland

Rebecca A Hughes

Head of Landscape & Restoration, Scottish Natural Heritage, 2 Anderson Place, Edinburgh EH6 5NP

Abstract

Identifying landscape sensitivity through landscape character assessment is a means of monitoring fragmentation and change in landscapes. This paper uses the results of one recent study to demonstrate how this cam be achieved constructively through the formation of guidelines for landscape enhancement and conservation of character. This study is that of Banff and Buchan in North East Scotland carried out in the Autumn of 1993.

Introduction

An appraisal of how the landscape of Banff & Buchan (Cobham Resource Consultants, 1994) has changed in the recent past is an important aspect of this paper. Anticipating future changes, and how these may affect the landscape character and quality of the District, is also necessary in order to formulate strategic guidelines for influencing and accommodating that change. Not all future changes will be detrimental, and some will present opportunities for landscape enhancement and increasing diversity throughout the District. Information on possible future change was obtained from the National Countryside Monitoring Scheme, Scottish Office Agriculture and Fisheries Department statistics, and through discussions with local authorities, the Forestry Authority, local landowners, Scottish Natural Heritage and the Scottish Agricultural Colleges.

Landscape Change This Century

Much of Banff & Buchan has been important for agriculture since early times. Although there is some historical evidence of extensive woodlands and forestry covering the District in the distant past, much of the land has, since the sixteenth or seventeenth centuries at least, been cultivated or under pasture. Today, broad expanses of intensively farmed land are occasionally broken up by pockets of policy plantings on estate lands.

In the western part of the District, some change has occurred between 1940 and 1970 through the planting of coniferous forest on hill land, with 700 km^2 under commercial forestry in the whole of the District by 1970. Since the 1970's however, little new coniferous planting appears to have taken place in the District as a whole; this is possibly a reflection of the price of agricultural land and the tax changes of 1986, inhibiting potential investors. Woodland Grant Scheme applications over the past five years have increased, but according to the Forestry Authority the majority are for less than 5 hectares of planting.

Changes in agriculture have initiated the greatest impacts on the landscape of Banff & Buchan, particularly in the past forty years. Fiscal and strategic policies aimed at increasing yields, improvements in farming technology, and fluctuations in economic markets, have all contributed to this change.

FRAGMENTATION IN AGRICULTURAL LANDSCAPES

The pattern of small farm units probably diminished after the agricultural revolution of the eighteenth and nineteenth centuries, with the population being encouraged to move to new planned settlements, thus clearing the land for large scale agriculture. The domination of a few large estates during this period resulted in a high proportion of tenant farmers throughout the District. By the early twentieth century many of these large estates became unprofitable and were dismantled and the land sold. The agricultural collapses during the early 1970's and poor harvest in 1985-87 resulted in many farms being sold, often to incomers, and a general amalgamation of units. Today a pattern of larger scale farm units in the east, and smaller farms in the more upland and less favoured areas to the west, has emerged.

Agricultural improvements since 1940 have resulted in 95% of heather moorland being converted to improved/unimproved grazing land and occasional commercial forestry, leaving just 1.59 km^2 by 1970. Much of this moorland would have provided rich habitats and been a significant feature on the upland fringe areas, increasing the visual diversity of the landscape. Likewise, the intensification of agriculture, bringing about new cultivation techniques and drainage schemes has had a dramatic effect on the extent and quality of lowland mosses which were, until comparatively recently, a feature of the landscape of the District. Some areas of lowland moss have been, and still are, cut for both domestic and commercial use, and although this has not occurred on a large scale, poor cutting techniques have led to localised degradation of mosses.

Hedgerows are present in some parts of the district, with many localities displaying a variety of species: beech and holly on lands owned by estates; gorse in the western uplands and more marginal areas close to the northern coastline. In many areas, farm boundaries were traditionally marked by ditches and stone walls. As farms have amalgamated and changed hands, and field sizes have increased, hedgerows have been removed - almost a third have gone since 1940 and walls have been left to decline. Fences have often replaced lost hedgerows and these have diminished diversity and affected landscape character in many areas.

Road-widening schemes and other developments have also brought about further loss of hedgerows, avenues and individual trees, with many of these distinctive features not being appreciated until they have disappeared. A notable exception has been the fight to retain a beech avenue near Mintlaw resulting in the formation of the Buchan Countryside Group. General neglect has resulted in the decline of distinctive beech hedgerows, while the management of road verges, and roadside hedgerows by both roads authorities and farmers has, in many areas, reduced the habitat value of these semi-natural features.

The Common Agricultural Policy (CAP), however, has brought about the most significant recent change to the landscape. Farmers now need to respond quickly to changes in subsidy by growing new crops. This has introduced significant changes to the appearance of the Banff & Buchan countryside, with crops such as oilseed rape being the most obvious. The recent introduction of the "set-aside" scheme has also initiated a strong visual impact on the landscape, with an estimated 1200 farmers in Banff & Buchan taking up a scheme with a set-aside component. In the summer months, the appearance of fields under set-aside, will have a strong visual contrast to adjacent, intensively farmed land. The current regulations governing the "set-aside" scheme inhibit long-term conservation projects and woodland planting, thus reducing the scope for positive enhancement of the landscape.

Current Initiatives

Current initiatives which contribute to landscape management and enhancement in Banff & Buchan include the Woodlands Grant Scheme (WGS) administered by the Forestry Authority, and the joint initiative by Grampian Regional Council, Banff and Buchan District Council and SNH called the East Grampian Conservation Project. This provides enhanced grant aid for small-scale countryside conservation projects below the minimum size requirement for WGS. Small-scale environmental work is also undertaken by the Buchan Countryside Group. This is, however, presently limited to the south-east of the District and mainly involves projects aimed at improving recreational facilities.

Policy and development controls are formulated by the local authorities. Areas of Special Landscape Value (ASLV) are designated along much of the coastal strip of the District and in the broad swathe of the Deveron valley. Within these designated ASLVs, the District Council policy is to "....give priority to maintaining the character of the landscape". Conservation areas have been designated for nineteen planned historic settlements and parts of Fraserburgh and Peterhead. In these designated areas "....new development will be carried out in a manner in keeping with the established character of the area". Advice on the siting, form and design of new building in the District is given in the draft Local Plan.

Grampian Regional Council has prepared a draft Forestry Strategy and has a policy of encouraging appropriate woodland cover and commercial forestry, as a means of diversifying land-use and employment opportunities in Banff & Buchan.

Potential Future Change

Continued changes of the CAP are likely to affect the rural economy and have a visual impact on the landscape. The initial five year set-aside scheme, now coming to an end, may be continued in the future, although many farmers have now taken up the Arable Land Payment Scheme, where 15% of arable land must be set-aside. Farm diversification schemes are generally limited within the District; an increase in outdoor pig raising is occurring at present, but few non-agricultural opportunities on set-aside land are likely under present policies.

The general decline in agricultural incomes recorded in the District may bring greater instability in land ownership and land-use. Many smaller units are presently having difficulties in adapting to policy changes, with larger farm holdings now being run as better full-scale businesses and generally managing to utilise subsidies. An increase in the size of farms is therefore likely in the future, as smaller units come on the market. There appears to be little evidence of an increase in small scale "crofting" units, generally popular with incomers, in the District, but this may become a trend in some areas if agricultural land prices fall further.

Positive management of agricultural land conservation and amenity is not presently a significant factor in Banff & Buchan. Although WGS applications and applications for grants under the East Grampian Conservation Scheme have increased, the majority are small-scale and taken up primarily by smallholders rather than full-time farmers. Forest cover is only likely to increase in the District if land prices drop to a more attractive level for private investors, as there would appear to be a reluctance on the part of farmers, at present, to diversify into woodland. Reasons for this reluctance would appear to centre on the attachment felt towards "hard-won" farmland, little cultural association with trees and possibly a lack of realistic financial incentives.

Although housing demand is not great in the District, there are likely to be continued applications for housing in the south and east of the District, within commuting distance of Aberdeen. This type of housing will, in the main, be concentrated within existing settlements. More important for the potential impact on the landscape, however, is the constant local demand for rural housing spread throughout the District.

Pressures for industrial development may not be significant in scale, but nevertheless some types of development may bring substantial changes to the countryside. Grampian Regional Council's Structure Plan supports further mineral extraction, mainly of sand and gravel deposits in the District. The District Council has a three-tier plan for mineral extraction, with a presumption against development in the majority of the coastal area, within designated landscapes and on high grade agricultural land.

The development of wind power generation facilities is also supported in principle by the Regional Council. The advent of the Scottish Renewables Order (SRO) is likely to produce interest in Banff & Buchan from potential developers of windfarms, because of the area's high wind speeds.

Overview of Issues of Landscape Sensitivity

The previous section considered the effect of past change on the present appearance of the landscape on Banff & Buchan. Potential pressures for change and development in the landscape were also assessed and a number of interrelated issues arising from these pressures can be identified. The key issues affecting the landscape of Banff & Buchan can be summarised as:

- **The changing character of the agricultural landscape.**
  The visual impacts of set-aside land, and the inhibiting effect of the present policy on positive schemes aimed at improving landscape diversity, are important issues in Banff & Buchan, where uptake is relatively high, future agricultural policy changes will be vital in determining the appearance of the countryside in the long term.

- **Continued threats to semi-natural habitats throughout the District.**
  It has been shown that features such as hedgerows, heather moorland and lowland mosses have been dramatically reduced in the past. Threats continue to be posed to the remaining mosses, areas of sand dunes, rivers and some areas of semi-natural woodland, as a result of pressures for peat extraction, coastal recreation, continuing trends towards agricultural intensification, and over-maturity and/or lack of management of policy woodlands.

- **A lack of landscape diversity and distinctiveness within intensively farmed land in many parts of the District.**
  As farm units have increased in size, and agriculture has become more intensive, hedgerows, woodland and semi-natural vegetation have diminished, leaving much of the Banff & Buchan landscape uniform in appearance.

- **A gradual deterioration and erosion of existing distinctive features.**
  Features which make a great contribution to landscape character in some parts of Banff & Buchan include stone walls, beech avenues and hedges, and mixed species roundels. Many of these are under threat of decline due to a lack of management, or removal for development. Distinctive architectural details, for example, traditional colour finishes and indigenous building materials, are not being used in many new buildings, thus diminishing landscape distinctiveness in some areas.

- **The impacts of present and potential new housing development on the fringes of existing settlement and within the countryside.**
  Poorly sited and designed housing detracts from and diminishes landscape quality. The demand for new housing in rural locations throughout Banff & Buchan is likely to continue and there is a need for any further development to reflect more closely local traditions in settlement layout and built form.

- **The potential pressure for new rural development.**
  The most significant future development pressure is likely to be for windfarms and large-scale forestry. The potential for increasing mineral extraction has also been identified in the Structure and Local Plans, but this is likely to be on a relatively small scale, given the distance from major markets.

The aim of the strategy is to address these issues through the encouragement of new initiatives and the preparation of guidelines.

Strategic Guidelines

The main objectives in a landscape strategy for Banff & Buchan would be the **conservation** of landscape features which make a positive contribution to landscape character, together with the **enhancement** of those landscapes which lack diversity and distinctiveness, or where there has been an erosion of character. In addition to these objectives, it is recognised that potential development needs to be strategically planned and sensitively accommodated without detriment to landscape character.

The Conservation of Existing Landscape Features

The distinctive landscape features of Banff & Buchan include natural elements such as the coastal landscapes, rivers and the lowland peat mosses. Man-made features include estate plantings, hedgerows and stone dykes.

While the protection of much of the coast of Banff & Buchan from large-scale development and detrimental change has largely been assured through the planning policies of the District Council, the gradual decline of other landscape features, such as estate woodlands, distinctive hill-top copses, hedgerows and avenues is occurring through a lack of management and restocking. The instigation of a programme of management and restocking is therefore vital to ensure that these features are not ultimately lost. In addition, the conservation of stone dykes need to be encouraged in areas such as the northern coastal strip, where they are sufficiently intact to make a valuable contribution of landscape character.

Lowland mosses, although considerably reduced in area since 1940, are an interesting feature in parts of the District. It is likely that an increase in the designation of some of these mosses will occur, following a current review by SNH. More sympathetic farming methods need to be encouraged at the perimeter of these mosses to ensure that nutrient run-off and changes to drainage do not affect their condition in the long term. Where peat is extracted from mosses, traditional extraction methods need to be used to minimise the impact on the habitat value of the mosses, and to conserve the visual characteristics of these valuable features.

FRAGMENTATION IN AGRICULTURAL LANDSCAPES

Opportunities for New Planting

The likely future increase in woodland cover in the District can be seen as an opportunity, as sensitively located and designed woodlands and forests have the potential to enhance the landscape by increasing diversity. The promotion of forestry is viewed by Grampian Regional Council as a way forward in counteracting agricultural decline and encouraging new landuses and processing industries into the District. Much of Banff & Buchan would be well suited to growing coniferous species on a commercial scale. Broadleaved species are unlikely to attain a high economic value, due to climatic conditions, although they would bring important amenity benefits.

The planting of new woodlands and forests could provide opportunities for reinforcing existing landscape character and for enhancement in areas where the landscape lacks diversity. The location and scale of new planting will, however, need to be carefully considered in order to protect and conserve existing landscape features and to ensure that the most effective use is made of new woodlands in terms of attaining visual enhancement and improving nature conservation interest. In addition the design of the forest edge will be particularly important in these flatter landscapes and should include a high proportion of broadleaved species.

In the more undulating agricultural heartlands small-to-medium scale mixed species woodlands and shelterbelts could be planted, linking with fragments of estate plantings and with hedgerows and copses occurring in localised areas. This would not only reinforce the existing landscape character, but also optimise the visual benefits of the new woodland.

Towards the west of the District, where coniferous woodland are more evident in the upland landscape, new planting should be located to visually link with these, thus ameliorating problems of poor scale or design. Some existing woodlands are poorly scaled and lack diversity in this area, and extended planting of both coniferous and broadleaved species could rectify design problems and make these woodlands positive landscape features.

Farm Conservation Schemes

As much of the Banff & Buchan landscape is under intensive agriculture, farm-based conservation schemes offer a major opportunity for introducing diversity into landscapes requiring some enhancement. Improvements to the current set-aside policy could encourage larger-scale and more widespread conservation schemes than those currently being undertaken. This could provide opportunities for farmers to establish a range of habitats; for example broadleaved woodlands, species-rich grasslands, wetlands and hedgerows. The visual significance and conservation value of these new features could be maximised by linking with existing hedgerows, riverbanks, and semi-natural vegetation, and by integrating conservation proposals for adjoining farms where possible.

Priority needs to be given in farm conservation schemes to the management and restocking of existing woodlands and avenues, where these make a positive contribution to the landscape. The planting of new, medium-scale, broadleaved and mixed woodlands would have a dramatic effect on the landscape in the future and should also be given an important emphasis in any strategy. The main water courses of the District could be enhanced further through waterside habitat creation on land taken out of production, and should also be seen as a priority within the promotion of farm conservation schemes.

FRAGMENTATION IN AGRICULTURAL LANDSCAPES

Housing in the Countryside

Future pressure for new housing is likely to continue to be concentrated in the countryside, although limited housing development will be located in some of the towns and smaller settlements.

New housing needs to be sensitively located, corresponding to the traditional patterns of settlement found in each character area.

Many settlements in Banff & Buchan make an important contribution to landscape quality, serving as focal points with views across the countryside. These views need not be compromised, and new development should therefore be restricted to sites where landform, and existing and new planting, offer opportunities for screening. In the countryside, the District Council's planning guidance should be enforced, with ribbon development and prominent hilltop sites being avoided. Integration could be achieved by locating new buildings close to existing buildings, shelterbelts, hedgerows and trees, where possible, or by planting native trees to replicate the copses traditionally planted for shelter around farmsteads. Coniferous hedges and fencing used for the plot boundaries of many new housing developments can be very intrusive in rural locations.

The detailed design of new building should reflect the distinctive forms and materials used locally. The simple forms of single and one-and-a-half storey farmsteads and village houses should be adhered to, although modern yet sensitive interpretations of these forms can provide opportunities to create attractive living spaces. The traditional proportions and style of windows and doors present examples of good design, while the building materials used should correspond to the differences found in the District, for example rendered wall finishes in and around many coastal villages, and the use of stone detailing in some inland areas. Opportunities for accentuating local distinctiveness should be encouraged and this might include the use of traditional colour details on new farm buildings and new stone-dykes on roadside boundaries.

Other Built Development

Potential large-scale new development in the countryside is likely to be confined to windfarms, mineral extraction and limited extension to the St Fergus Gas Terminal. While it is expected that the latter two developments will be limited in extent and have a mainly local visual significance, the development of windfarms would have major visual impacts on the countryside.

Large-scale windfarm development would be inappropriate in Banff & Buchan, the overall landscape character of this medium-scale, settled and strongly agricultural area being unable to visually accommodate large numbers of turbines. Smaller scale developments of up to twenty turbines could, however, be accommodated in some parts of the District with care. Any such development should be directed away from landscapes being settled by small scale farms and clustered villages and being more "domestic" in character, and thus highly sensitive to the intrusion of large-scale elements. The enticing natural features and seascapes of the coastal landscapes would also be diminished by such a large-scale feature. Overhead transmission lines providing connections between windfarms and the national grid could potentially cause significant damage in the generally uncluttered, wide and open landscapes of Banff & Buchan.

Any additional radio masts, should be resisted, a "saturation" level having already been reached in some areas in terms of their adverse effect on landscape quality.

Mechanisms for Change

Planning authorities, land-use agencies and conservation bodies all have an important role to play in devising mechanisms which will influence the future appearance of the Banff & Buchan landscape. The main mechanisms operating at present fall mainly under development control, and countryside management and enhancement. Future CAP reforms, favouring long-term conservation schemes on land taken out of production, offer major scope for diversification and enrichment of the agricultural landscapes which form much of the District. Mechanisms for creating new landscapes will be important for Banff & Buchan, and there would appear to be scope for partnership initiatives to be instigated, utilising limited resources and directing more wide reaching enhancement projects.

The landscape of Banff & Buchan has been found to display a range of different characters and have many attractive and distinctive landscape features. In parts of the District, however, intensive farming has diminished landscape diversity and some new development, although limited in scale, has also adversely affected landscape quality. A broad vision for the future of Banff & Buchan landscape needs to:

- maintain and enhance the traditional farm, estate and village landscapes, whose origins lie mainly in the eighteenth and nineteenth century settlement of the District;
  conserve the special natural character of the coastal area and traditional settlement patterns of the District of directing new development to build on existing visual and architectural character;

- sensitively site any new large-scale development so as to minimise landscape and visual impacts;

- encourage landscape enhancement through the expansion of natural habitats on farmland and through the creation of a more wooded landscape, particularly in the agricultural heartlands.

References

Cobham Resource Consultants (1994) *Banf and Buchan Landscape Assessment*. Unpublished Contract Report SNH/031A/94/LRB, Scottish Natural Heritage, Edinburgh.•

Developing a sustainable landscape : the dilemma of the Llŷn Peninsula.

E.E.Thomas

Department of Environmental Planning and Management, Coleg Normal, Bangor, Gwynedd, Britain. LL57 2PX

Abstract

In view of the recent pressure for change in farming methods away from intensive land use towards an approach which favours the environment, a field study of the farming population of the Llŷn Peninsula was carried out in 1990. Factors such as farming intensification and afforestation which have been reported to influence habitat change (Howe *et al.*, 1990 and Buse, 1992) were investigated. These were considered to be particularly important as the study area is designated as an Area of Outstanding Natural Beauty and is classified as a landscape of great conservation value. The Llŷn Peninsula has also been designated an Environmentally Sensitive Area with many Sites of Special Scientific Interest specifically areas of wetland and coastal heath.

Introduction

The paper presented explores the meaning of sustainability of landscapes, specifically the Llŷn Peninsula, North Wales as an example of a highly designated agricultural landscape. The environmental and socio-economic context of the process leading to designation of the area as an Environmentally Sensitive Area (E.S.A.) is summarised as is the farming response to the scheme. The contribution the scheme could potentially make to the sustainability of the area's farming and wider community is also discussed.

Sustainability on the landscape scale

In the designated areas of rural Britain, in coastal as well as upland areas, the development of a viable population and economy is a fine balance between sustaining the scenic nature of the environment, whilst also supporting the communities which populate these areas. This must be considered alongside the need for nurturing the resident culture which is often one of the strongest forces pulling people to visit and to settle in these areas.

In this context, the term sustainability of landscape is applied to address issues of landscape quality and habitat diversity on the local or human community scale. Diversity of habitats is taken as an indication of the sustainability of particular ecosystems at this scale. Thus developing (or maintaining) a landscape system which allows humans to change to "meet the needs of the present without compromising the ability of future generations to meet their own needs" (World Commission on Environment and Development, 1989) would, in this sense, apply to the maintenance of landscape and habitat diversity via state support administered through the various government and non-government agencies. Integration of economic and environmental concerns occurred in some degree during the preparation of the Department of Environment (DoE) White paper "This Common Inheritance" wherein simple statements described the recently crystallised strategy which included landscape conservation and improvement, and also the conservation of the diversity of Britain's Wildlife by protecting habitats (DoE, 1990, pg.96).

Discussions on sustainable rural development (see Midmore, 1994) applied to landscape issues would suggest that the resource, in this case landscape quality and habitat diversity need to be maintained in terms of potential to generate future income. The potential income relates in this case to the "owners" of the valued areas and also the wider economic community it both directly and indirectly supports for example through tourism alongside traditional agricultural activity.

A different strand in the sustainability debate within the agriculture industry is the need to decrease farm inputs or to de-intensify be it on a small proportion of the farm (for example set aside) or on a whole farm scale such as the E.S.A. scheme applied to Llŷn and many other areas. In developing the sustainability theme such a statement as "Natural capital is now in shorter supply than man made and hence becomes the limiting factor" (Meldon, 1993), strongly indicate that for western agriculture to move into line with such thinking, recent policy movements have to be further researched and developed especially farmers and local community response to such initiatives. From recent work in this field the burden of responsibility of bringing British agriculture into the realms of sustainability is mainly being borne by the more marginal areas - set aside has received a fool's welcome but not in most cases as a whole farm strategy. It seems therefore that post-Rio directives from Brussels are applied in contradiction to the ideal of sustainability - namely that highly productive land is exploited until it is exhausted for the benefit of capital accumulation on very few large farms. The explanation is that production on marginal land is lowered to decrease the total area of land in production in Britain, thus increasing sustainability seems to be not quite right. Local people are, it could be argued, included in the decision making process through the concept of subsidiarity as they are voluntary participants in these schemes and as such they are able to opt in or out of participation in EU Environmental Action Programmes such as the *Towards Sustainability Programme* (European Union Council of Ministers, 1992). In this way the decision to participate in environmental and developmental projects is brought as close to the citizen as possible thus encouraging increased responsibility for agricultural change using a bottom up approach. The application of such policy measures will be discussed by taking the Llŷn Peninsula as an example of an area which is highly designated in terms of both nature and landscape conservation.

The loss of habitat diversity

Areas of the periphery such as West Wales, N.W. Scotland, West England and Ireland have, as a result of their geographies been areas of low tech farming (not necessarily though low input of labour in most areas). In many of these areas agricultural development towards more intensive systems has occurred since the mid 1970's in response to LFA subsidies to the detriment of habitat diversity and landscape quality (see for example Barr *et al.* 1983; Countryside Commission, 1990 (a); MacEwen and Sinclair, 1983). This has been the case in both Wales (e.g. Buse, 1992, Wathern *et al.*, 1988) the rest of Britain (e.g. Bowers and Cheshire, 1983; Lowe *et al.*, 1986).

In Llŷn for example, there has been an overall increase of approximately 5 per cent in land used for agriculture between 1975 and 1988 (calculated from W.O.A.D census data). However, this trend is not uniform throughout the peninsula as some of the more urban parishes and those susceptible to pressure from residential development, have shown a decrease in agricultural area in the same period. These losses have, however, been compensated for by the increased areas farmed the more upland parishes where agricultural land area increased by up to 50% of the original farmed area, respectively. The increased area of land brought into agricultural cultivation may also be associated with a contraction in the semi-natural, present on the original farmed area, through cultivation, as indicated by the agricultural parish censuses (Thomas, 1993). Agricultural land may be seen as slowly creeping upland to be replaced in the lowland areas by development - mainly residential.

Agricultural subsidies seem to have exacerbated the situation as parishes with L.F.A. (a) designation, and thus the highest rate of headage payment for Hill Livestock Allowances, experienced a fall of 22% in the original area of rough grazing occurred during this period. This trend was seen throughout the peninsula, with an associated rise in the original crop and fallow area (including improved grassland and arable). These rates of loss are comparable with those identified by Buse for the same areas between 1972 and 1985. From this it can be concluded that the rate of loss was relatively constant throughout the fifteen year period. The proportion of farmed land under woodland has also increased overall by 55% of the original area. This increase is mainly due to coniferous plantation. This may be due to the cost and investment required in planting broad leaved trees. There was also a reported low rate of uptake of woodland grant aid in Llŷn up to 1992 (less than one percent of those eligible (Welsh Office Agriculture Department, 1993)).

The overall trend within the lowland parishes in Llŷn is that of contracting agricultural land area, with the greatest decline in farmed areas occurring in the coastal parishes associated with tourism, under pressure from development. The least contraction of farmed land has occurred in parishes with the best agricultural land quality. The overall percentage change in farmed area suggests very strongly that agricultural land use is becoming concentrated on the most profitable land. This has created a polarisation of agriculturally wealthy parishes contrasting with agriculturally marginal parishes and also a trend of extensified land use in the coastal parishes based on the tourism industry (Cyngor Dosbarth Dwyfor, 1991,1988).

Loss of habitat diversity and fragmentation

Habitat change has been unusually well documented in the Llŷn Peninsula. The area has been surveyed during the first phase of the Nature Conservancy Council's (now part of the Countryside Council for Wales) vegetation survey. During the Nature Conservancy Council's (N.C.C.) survey, Llŷn was part of the North Wales "area of scarch" whereby new Sites of Special Scientific Interest were identified.

In the survey, unimproved grassland was reported as rare on the low-lying areas of Llŷn due to previous agricultural improvement (Howe *et al.*, 1990). It was expected that there would be some areas of unimproved haymeadow ("gweirglodd") present. In this region, farms have been less intensively developed, thus semi-natural rough areas on many of the holdings have been retained in an unmanaged state. Despite this, only very little (0.25%) of the semi-natural habitat was considered of primary ecological importance (due to its unimproved state). This was in the west of the E.S.A. (Howe *et al.*, 1990). Marshy grasslands occurred predominantly in the east of the peninsula. The main areas of semi-natural broadleaved woodland were lost before the sixteenth century. Of the woodland remaining in Llŷn in 1987-88, one third was coniferous. The areas of conservation interest encompassed two of the traditionally managed estates of Cefnamlwch and Bron Eifion. A large area of semi-natural, broadleaved woodland still exists, however, protected in the Coed Elenydd S.S.S.I.

The ecological interest of the coastal marshlands has been reduced by development in many locations, especially on the south coast. Important coastal heath now only occurs in patches in the extreme west of the peninsula. It is greatly overgrazed in many locations (Howe *et al.*, 1990). A large proportion of this habitat is designated as S.S.S.I. Wet heath habitat (or "Rhos", "Gwaun", "Wern") is also very restricted in distribution and is greatly fragmented (Howe *et al.*, 1990). The wetland mire habitats ("Cors" and "Mawndir") occurred mainly in the south and have been threatened by drainage. Many are of S.S.S.I. status and subsequent to the report a significant area of valley floor mire in central Llŷn has been purchased by Countryside Council for Wales and designated as a National Nature Reserve.

Designation of E.S.A. areas - the why and the wherefore?

Following the addition of Article 19 to the 1985 European Directive EC797/85, section 18 of the 1986 Agricultural Act, was introduced which allowed the Ministry of Agriculture, Fisheries and Food in the U.K. to identify areas where the intrinsic environmental characteristics of the area were being threatened by local farming activities. The first round of British E.S.A.s were designated in 1987. In January 1988, 7 more areas were designated in the second round, which included The Llŷn Peninsula. During 1992 a further round of E.S.A.s was proposed for designation in England and Wales schemes being applied in May and the Autumn of 1993.

Some of the characteristic features of British landscape typical of particular regions of farming tradition are captured in these designations such as the dry stone walls and field barns of the Dales, the species rich calcareous pastures of the Downs and the wilder scenery of the Peak Moorland. These schemes are an attempt to conserve agricultural landscapes created in the U.K. through the regionally differentiated evolution of the industry and to maintain these diverse, visually rich areas in the face of changing agricultural intensity.

In parts of Llŷn over 40% of the economically active population are employed in agriculture (Office of Population Census Survey, 1981). Any radical change in farming methods in Llŷn could be compounded, and would, significantly alter the landscape within a very short time scale e.g. through afforestation. Recent farming trends have, therefore, been a fundamental consideration for the development of a nature conservation strategy in Llŷn (Cyngor Dosbarth Dwyfor, 1993).

All these factors, coupled with the threat of heavy sheep grazing pressure on the rare coastal heathlands, led to the designation of Llŷn as an Environmentally Sensitive Area in the second round of designations in 1987-88. This was seen as a potential solution to the requirement for non site specific conservation of habitat. In their discussion of the implications of these habitat losses for bird populations in the U.K., the degree of change was clearly associated with the need for environmental policy formation (R.S.P.B., 1984). This was also the case in Llŷn where there are several sites of international significance for rare bird populations such as the Chough.

The E.S.A. of the Llŷn Peninsula was designated in January 1987 as being the whole of the north western peninsula of North Wales bound to the east by the Snowdonia National Park, totalling some 37,900 hectares. The specific features of the area considered to be worthy of protection were the earthbank-bound field patterns, stone or slate gate posts, areas of roughland, wetland and haymeadows, together with the preservation of more general features of the small scale landscape created by the local farming population.

The E.S.A. provided the farmer with a set rate at January 1st 1988 of £15 per hectare per year paid half yearly. In return for this the farmer signed a legally binding five year management agreement with W.O.A.D. for the whole farm. For some areas of land with specific wildlife value, for example, wetlands or roughland, a higher rate of payment at a set rate for the whole of the E.S.A. (at January 1st 1989) was paid at an additional £15 per hectare per year. In return, this land was submitted into a higher level of designation, Tier 2(I), for which there was more rigorous management conditions (see W.O.A.D., 1988).

In order to establish whether the E.S.A. schemes are actually working to slow down or to prevent change in the landscape features described previously, a landscape assessment of each E.S.A. has been carried out during the first year of designation. For example an assessment of the landscapes of both the Cambrian Mountains and Llŷn were carried out by Countryside Commission (1990 (b)) and by W.O.A.D. (1991), respectively. A socio-economic survey was also commissioned to investigate the effects of both E.S.A.s (W.O.A.D., 1992).

Response to the scheme

The process of designation of the area as an E.S.A. was welcomed by the conservation organisations. The E.S.A. scheme, acting alongside the already strict housing and development planning restrictions on the peninsula was seen as a means whereby landscape characteristics valued by the residents and by visitors to the area might be conserved, with the co-operation of the farming community.

By July 1993, 55% of the 960 eligible farm businesses had joined the scheme (W.O.A.D.,*pers com.*). From the results of an in depth survey of the Llŷn farming community shown in Table 1, it was found that the farmers most likely to join were farming relatively large holdings, were possibly pluriactive and from extended families. Poorer land quality was an obvious indicator of tier two membership.

Contribution of the E.S.A. schemes to sustainability in Llŷn

<u>Habitat diversity and landscape quality</u>

Successful protection of a sufficient area of representative vegetation types necessitates that sufficiently large non-fragmented areas of land be submitted into Tier 2, especially areas of the rarer types such as alkaline wetlands and semi-natural grazing areas.

Table 1 **Uptake of the Llŷn Environmentally Sensitive Area scheme (Thomas, 1993).**

	Proportion of sample	Area of land/ha
Non-members	54.4%	33.0%
Tier 1 only	25.6%	58.3%
Tier 1 and 2	20.05	8.7%

From a comparison of the landscape classification (W.O.A.D., 1991), it is evident that a reasonable degree of protection is afforded to most of the landscape types described. This is suggested by the rates of land entry into Tier 1 agreements throughout Llŷn. A comparison of uptake with the vegetation monitoring report (Howe *et. al,* 1990) shows that only semi-natural habitats in the parish of Boduan are sufficiently protected by Tier 2 designation. The most diverse areas occur in central Llŷn, where less than 1% of the sampled area was in Tier 2 of the scheme. A significant area of the extreme west and north east of the peninsula have been submitted into the scheme which probably covers roughland, composed of unimproved grassland, and some coastal heath. As landscape change brings with it a loss of species diversity and often lowers the scenic value of an area, a scheme with the objectives of habitat conservation is a positive move in policy formation towards sustainability in Llŷn.

Sustainability of Llŷn's farming system.

Through the process of reducing production, the concept of the extensification of farming also contributes towards general agricultural decline. The role of the farmer becomes that of landscape custodian once again. Farmers in the area are on the whole reasonably satisfied with this type of payment, (if rather suspicious of the obligations associated with the scheme). The income generated by the scheme to the average farmer in the sample was calculated to be approximately £1,300, (taking all members to have the same area in Tier 1 and Tier 2). The average net farm income in this area was gauged to be one of the lowest of all the proposed E.S.A.s at £5,000, (Countryside Commission, 1986). Therefore the E.S.A. scheme could,in theory from these calculations, provide the average farmer with an additional 25% extra to his/her normal net farm income. However, in reality most payments for land submitted into the scheme identified by the survey were much lower than this figure due to the great number of small farms the area. The scheme in some respect took advantage of a period of farming depression when very little opposition was expected to its conditions or to its objectives. Very low payments were offered for joining the scheme, subsequently tier 2 membership suffered.

Both U.K. and European policy for marginal areas seem to be encouraging decline through increased competition and specialisation. At the same time, as financial handouts are available to the same farmers "to do nothing", farmers in the more prosperous lowlands of Britain are still encouraged to "produce". The E.S.A. scheme may be viewed from various perspectives, ecologically it is an opportunity for Llŷn to move towards sustainability in terms of landscape (Tier 1) and perhaps less successfully for habitat diversity, given the more disappointing farming response to Tier 2. Politically there seems to be a hidden socio-economic agenda, based on the dependence of increasingly marginalised rural areas on the governing urban consumer core and the high input productive agricultural region - a situation which does not, within the time scale of generations and given the whim of policy makers, encourage much hope of lasting real sustainability.

References

Barr, C.J., Ball, D.F., Bunce, R.G.H., Risedale, H. and Whittaker, M. (1983) Landscape Changes in Britain. *Annual Report. Institute of Terrestrial Ecology*. Merlewood.

Bowers, J and Cheshire, P. (1983) *Agriculture, the Countryside and Land use - an Economic Critique*. Methuen.

Buse, A. (1992) Environmental Effects of Land Use Change, as Identified by Habitat Recording: a Case Study in the Llŷn Peninsula Wales. *Journal of Environmental Management,* **35:** 131-151.

Countryside Commission (1990)(a) *Changes in Landscape Features in England and Wales 1947-1980.* CCD44 Countryside Commission, Cheltenham.

Countryside Commission (1990)(b) *Cambrian Mountains Landscape Assessment.* CCP 293, Countryside Commission. Cheltenham.

Countryside Commission (1986) *Outline Proposals for E.S.A.in Britain.* Cheltenham.

Cyngor Dosbarth Dwyfor (1993) *Countryside Management Plan. Dwyfor District Council.* Pwllheli, Gwynedd.

Cyngor Dosbarth Dwyfor (1991) *Cynllun Lleol Dwyfor: Dogfen Drafod. Dwyfor District Council.* Pwllheli, Gwynedd.

Cyngor Dosbarth Dwyfor (1988) *Cynllun Lleol Dwyfor Wledig, Datganiad Ysgrifenedig* Dwyfor District Council, Pwllheli, Gwynedd.

Department of Environment (1990) *Our Common Inheritance.* HMSO London.

European Union Council of Ministers, (1992) *Towards Sustainability, A European Community Programme of Action*, Brussels.

Howe, E.A., Rimes, C.A. and Blackstock, T.H. (1990) *A Habitat Survey of the Llŷn Peninsula.* N.C.C.Wales, Bangor.

Lowe, P. *et al.*, (1986) *Countryside Conflicts. The Politics of Farming Forestry and Conservation.* Gower. Aldershot.

MacEwen, M. and Sinclair, G. (1983) *New Life for the Hills - policies for farming and conservation in the uplands.* Council for National Parks. London.

Meldon, J. (1993) The Environment and Development: Towards a Sustainable Future? *Geographical Viewpoint*, **21:** 57-66.

Midmore, P. (1994) *Notes towards a definition of Sustainable Rural Development.* Paper presented at the Wales Rural Forum, University of Wales Residential Centre, Gregynog.

O.P.C.S. (1981) *Population Census Data*, Central Government Statistical Office, HMSO London.

Royal Society for the Protection of Birds,(1984) *Hill farming and Birds: a survival plan.* Sandy, RSPB.

Thomas, E.E. (1993) *The Socio-economic Context of Agricultural and Ecological change in the Llyn Peninsula.* Ph.D. Thesis, University of Liverpool.

Wathern, P., Young, S.N., Brown, I.W. and Roberts, D.A. (1988) Recent Upland Land Use Change and Agricultural Policy in Clwyd, North Wales. *Applied Geography,* **8:** 147-163.

Welsh Office Agricultural Department (1992) *Socio-economic Monitoring Report for the Welsh E.S.A.s.* W.O.A.D. Cardiff.

Welsh Office Agricultural Department (1991) *The Llyn Peninsula E.S.A., Landscape Assessment for Monitoring.* W.O.A.D. Cardiff.

Welsh Office Agricultural Department (1990) *Ystadegau Amaethyddol Cymru* Rhif.12 1990, Swyddfa Gymreig, Caerdydd.

Welsh Office Agricultural Department (1988) *Llyn Peninsula E.S.A.* - Farmer's Information Pack. W.O.A.D. Cardiff.

World Commission on Environment and Development, (1989) *Our Common Future*, Oxford: Oxford University Press.

PART II:

FRAGMENTATION IMPACTS ON FAUNA

The effects of Habitat Fragmentation on Invertebrate Distributions

S.J. Falk

Herbert Art Gallery & Museum, Jordan Well, Coventry CV1 5RW, United Kingdom

Abstract

Different invertebrate groups have varied in their response to changes in the British countryside brought about by agricultural intensification, changing agricultural practices, urbanisation, etc. Substantial declines are frequent within groups such as butterflies and aculeate Hymenoptera. These have usually led to impoverished zones (for such faunas) corresponding with areas of intensively managed agricultural landscapes. Reinstating a high diversity of such faunas into these landscapes will prove very difficult. Other insect groups have proven more resilient and can sustain fairly impressive levels of diversity in agricultural landscapes. This is usually related to factors such as high mobility, the ability of many species to utilise small habitat fragments, and the ability of many species to utilise temporary and non-ancient features. Farmers could promote such faunas relatively easily by protecting and/or creating a variety of habitat features.

Introduction

The last 2 decades have seen a major increase in the availability of distributional data for invertebrates. These include national atlases published by organisations such as ITE, county atlases from local record centres and taxonomic group reviews from NCC and JNCC. These have greatly improved our knowledge of how invertebrates are coping with changes in our countryside. Responses differ between groups. This paper provides a summary of the impact of changing land use on 4 insect groups - butterflies, bumblebees, solitary bees & wasps and flies, and briefly surveys some of the factors that may have rendered these groups vulnerable or relatively resilient to such changes.

Butterflies

By virtue of their popularity, butterflies have been particularly well studied. Most of our species have been subject to at least 3 lines of research relevant to this theme.

i) Mapping at national and local levels to produce atlases showing current distribution and former range. The most recently published national maps are to be found in Emmet & Heath (1990).

ii) Monitoring relative changes in the frequency of adults throughout England, Wales and Scotland by the Butterfly Monitoring Scheme. This important work helps us to distinguish short term cycles of frequency, which are operating nationally or regionally, from serious long term declines and site-related phenomena (Pollard & Yates, 1993).

iii) Detailed autecological studies investigating life history, ecology, adult mobility etc.

These 3 lines of study reveal that many butterflies have been adversely affected by changes to the British countryside over the past 150 years, and especially since the last war. Of our 59 recorded resident species, 4 are now extinct, and of the remainder, three-quarters have declined substantially and 6 are threatened with extinction (Warren, 1990). The ranges of many species have contracted markedly and the number of sites supporting unusually rich assemblages is much reduced. Such changes have taken place almost as rapidly on protected sites such as SSSI's and National Parks as unprotected areas (Thomas, 1984). Species associated with traditionally grazed unimproved grassland and open areas within woodland have been particularly affected.

Why are some butterflies vulnerable to habitat fragmentation?

i) Exacting requirements for egg laying and larval development, usually related to the need for certain types of vegetation structure. The existence of such conditions has tended to rely upon traditional habitat management, often augmented by factors such as rabbit grazing. Wholesale habitat destruction, changes in agricultural and silvicultural practices, and myxomatosis have resulted in a major reduction of habitats suitable for scarcer butterflies (eg Warren, 1990).

ii) Low powers of dispersal, which prevent populations from colonising vacant habitat nearby. Dispersal distances can be as low as 1 km in species such as the Silver-studded Blue (Thomas *et al.*, 1992).

iii) Spatially dynamic populations. These are populations that move about a site in response to changing conditions bought about by natural successional cycles, coppicing cycles, and year to year inconsistencies in management. Ideal conditions may last for only a few years in one place. This is not a problem providing that: ideal conditions are constantly being created elsewhere; there is always sufficient breeding habitat for viable populations; vacant niches suitable for colonisation are within easy reach. Unfortunately, habitat fragmentation tends to act against all of these criteria.

Metapopulations and sedentary butterflies

Several of our scarcest butterflies have been shown to survive today as groups of loosely affiliated populations (metapopulations). These occupy a variable number of relatively large, non-isolated habitat patches, either within a single large site or a number of closely approximated sites (Thomas *et al.*, 1992). Smaller habitat fragments and more isolated patches are usually not colonised, and the overall pattern of distribution resembles that predicted by Island Biogeography Theory. Providing habitat patches are relatively close, these metapopulations have the ability to colonise vacant niches as they appear, thus cancelling out the effects of local extinction from individual habitat patches. The rate of turnover (extinction and colonisation) will be highest in dynamic habitats such as heathland, certain sorts of grassland and actively coppiced woodland.

The survival of metapopulations such as these depends upon reasonable quantities of suitable habitat being available within relatively large areas of biotope. The transient nature of individual populations means that metapopulation systems are an effective way of utilising the habitat within an area and provide much greater resilience against factors such as bad weather, periods of mis-management etc. Conservation of such species should attempt to preserve metapopulation systems and this will rely partly upon reducing or reversing habitat fragmentation, and reinstating certain forms of management.

Promoting commoner butterflies in farmland

Many of our commoner butterflies are able to cope with fairly intensive agricultural settings, be it at lower densities than in the past. There is much scope for increasing the numbers of these by sympathetic management of field margins, hedgerows and unimproved grassland. Habitat creation in the form of small farm woodlands and long term set-aside may also be beneficial to some species. The importance of features such as disused railway cuttings, old quarries and various forms of wildlife corridors within farmland should also be stressed to farmers. In many counties, these provide major refuges for both scarce and common species, but are widely threatened. There are a number of projects currently studying butterflies in farmland and Butterfly Conservation has recently published a book aimed at advising land owners (Dunbar, 1994).

Bumblebees

In the last 40 years, 6 of our 19 species have shown substantial declines in southern Britain (Alford, 1980). Several further species are less frequent there now than at the beginning of the century and two species are presumed extinct. The loss of bumblebee diversity (involving both the decline of individual species and a loss of localities with 8 or more species) has been particularly marked in arable, 'planned' countryside of the Midlands and southern England, resulting in a Central Impoverished Zone with a fauna largely reduced to the 7 commonest British species (Williams, 1982, 1986). Even well known nature conservation sites such as Wicken Fen have been victim to this process.

Patterns of retreat vary between the 6 highly declined species, but most are now found in coastal areas and/or a patchy scattering of inland sites mostly outside the Impoverished Zone. The declines have also been accompanied in some species by a marked contraction in the variety of habitats utilised. Thus *B.distinguendus* has retreated almost entirely to coastal moorland and grassland in the far north of Scotland, *B.subterraneus* and *B.humilis* persist mostly in southern coastal habitats and *B.soroeensis* remains widespread on heaths and moors in the north and west of Britain. *B.ruderatus* and *B.sylvarum* survive in a variety of habitats, but all the declined species have difficulties coping with modern intensive agricultural landscapes, despite the fact that it was agricultural landscapes of pre-war years that permitted their presence in much of central and southern England.

This change in fortune by bumblebees suggests that they are useful indicators of the 'damage' caused by agricultural intensification. It is the prevailing opinion both here and abroad (where declines have also been noted) that the loss of good quality foraging habitat through changing land use is the main cause of these declines.

Some ecological characteristics of bumblebees

The foraging habits and other ecological characteristics of British bumblebees have received a good deal of attention (eg Alford, 1975, Fussell & Corbet, 1992, Williams, 1986, 1988, 1989). Understanding the following characteristics of bumblebees is crucial when formulating plans to conserve them.

i) Most British bumblebees can utilise a wide variety of flowers in a range of habitats. The ecological biases exhibited by species are related to tongue-length, behavioural traits and phenology, but much overlap of habitat and flower usage occurs between species.

ii) There is a heavy dependence upon perennial flowers growing in established habitats of moderate to high floristic diversity. This is particularly marked in the declined species which in the past tended to occur inland within landscapes that provided much unimproved permanent grassland.

iii) Several of the declined species appear to fall close to the edges of their natural ranges within Britain. Such species may be particularly dependent upon sites that permit efficient foraging. Such sites need to be large and to provide sufficient densities of suitable flowers throughout the foraging season.

iv) Population units are based on colonies and viable populations based on these make far greater demands than viable populations of most non-social flower-visiting insects. This results in a requirement for large tracts of suitable foraging habitat. Today, most scarce species persist in coastal areas characterised by large expanses of semi-natural habitat. Some of these species may be surviving as discrete metapopulations based on certain regional landscapes.

v) Observations at Dungeness, Kent (Williams, 1989) and Kenfig Burrows, Glamorganshire (Falk, *pers.obs.*), which are 2 of Britain's top bumblebee sites, seem to show that high densities of certain flowers are critical in preserving the presence of scarce species and diverse bumblebee faunas. These flowers, which include *Echium vulgare, Teucrium scorodonia* and *Cirsium vulgare,* provide high nectar yields and can be utilised by many of the bumblebee species at a site. The reduced density of such plants at Shoreham, also in Kent, and Merthyr Mawr Warren, near Kenfig Burrows, seem to be major factors why these latter sites support fewer species.

Can bumblebee diversity be restored in agricultural landscapes?

From what is now known about bumblebee ecology it seems unlikely that the scarcer species can be fully restored to their former range. The sorts of agricultural habitats they most relied upon were of an established, often ancient origin that would be difficult to recreate. Enhancing the population levels of the commoner species is more feasible. These already cope with intensive agricultural settings by utilising field margins and other established features within farmland. Their populations will almost certainly respond to the increased provision of foraging habitat. However, their preference for perennial flowers means that short term set-aside cycles are probably only of limited use and need to be lengthened (S.A. Corbet-*pers.comm.*). A major priority for the conservation of bumblebees today should be the identification and preservation of modern populations of the declined species and sites with high bumblebee diversity. These could be used as nuclei for bumblebee enhancement zones incorporating semi-natural habitats with carefully managed agricultural areas.

Solitary bees and wasps

This group also contains a large proportion of declined species, though much more recording is required to clarify the true situation. At present, approximately a third of the species are accorded with Red Data Book status, a further third with Nationally Scarce status and about 5% of the fauna is regarded as extinct (Falk, 1991a). These figures are higher than any other insect group. Many of the declined species once existed widely in agricultural landscapes of southern England, particularly where lighter soils prevail. The process of impoverishment in Bedfordshire is described by Chambers (1971) and that for the once excellent Hampstead Heath by Guichard and Yarrow (1948).

Ground-nesting species have been most badly affected which is perhaps not surprising as ground vegetation conditions and their associated microclimates are one of the greatest casualties of agricultural intensification. Dead wood- and stem-nesting species have fared much better by comparison and many such species seem to have increased in frequency over recent years. This may be a response to the large quantities of dead wood generated by Dutch Elm disease and the development of scrub and secondary woodland following the onset of myxomatosis in rabbits.

Modern refugia of solitary bees and wasps.

A number of these can be identified in Britain within the patterns of decline of formerly widespread species. Such species may now be surviving in one or several of them.

i) Coastal areas with habitats such as soft rock cliffs, extensive unimproved headlands and sand dunes. Such areas provide a good mosaic of conditions, including the bare earth required by ground-nesting species, and a variety of herbage types and scrub for foraging or hunting. Most crucially, these conditions are partially self-sustaining through natural erosion processes and exposure, freeing them somewhat from the whims of agriculture. Some declined species may now be entirely restricted to coastal sites.

ii) South-west and South-east England. Counties such as Cornwall, Devon and Kent are characterised by long stretches of varied coastline and a variety of habitats inland that are suited to aculeates. The species diversity of such counties remains relatively high. One explanation for this is that coastal areas act as a source of colonisation into vacant inland niches. In counties such as Devon, the characteristic hedge-banked landscapes are also important. These provide a huge resource of earthy, flowery slopes arranged in every orientation, plus hedgerows rich in flowering shrubs and trees. Such conditions suit mining bees in particular, and a number of our scarcest species persist in such places. Few species fall exclusively into this category, but tend to combine it with i) and iii).

iii) Semi-natural and disturbed habitats in 'inland' central and southern England. Some species still persist in heathland and downland districts within fragments of remaining semi-natural habitat. In many instances, it is the old quarries, railway cuttings and possibly motorway verges of such areas that now support the best assemblages. Urban landscapes have also become much more important, particularly in industrial towns and cities which have large quantities of industrial waste ground. In vice-county Warwickshire, the top aculeate site to date is a former spoilheap in east Coventry. Many scarce species combine this category with i) and ii).

<u>The need for habitat mosaics</u>

Perhaps the strongest message that can be derived from the decline of solitary aculeates is the need to preserve habitat mosaics. Most species have complex life histories with multiple requirements. Mining bees for instance need specific locations in which to nest, pollen sources and nectar sources, and possibly specific features for male swarming. In double-brooded species the floral demands are doubled. Non-intensive agricultural settings tend to be floristically and structurally diverse, thus supporting a much more complex array of habitat combinations than intensively-farmed landscapes. There is a need to examine in detail the manner in which combinations and mosaics of habitats are affected by agricultural intensification (as opposed to simply recording the net losses of habitat).

Flies

A number of fly groups have been subject to provisional atlases and several county atlases have been produced for hoverflies, reflecting the popularity of this group. Declines are apparent for groups such as horseflies, also species such as the robber fly *Asilus crabroniformis* (Drake, 1991). This probably relates to the loss of damp unimproved pasture and heathland. Many further declines are likely to exist within species utilising established wetlands, ancient woodland, ancient trees etc. However, as a generalisation, flies appear to be considerably more resilient than the previous groups and many decline less rapidly in response to habitat loss. The proportion of species falling into the various rarity categories is much lower (Falk, 1991b) and many species formerly furnished with Red Data Book or Nationally Scarce status are being downgraded as a more complete picture of their statuses emerges. This is especially apparent for species that were previously regarded as strict indicators of high quality ancient wetlands and ancient woodlands.

Intensive recording of many fly families in my own county (Warwickshire) is demonstrating the impressive diversity of species surviving in fragments of semi-natural habitats, despite the relatively intensive levels of agriculture and low coverage of semi-natural habitats to be found there. Often the quality and precise character of habitat is crucial, but small and isolated fragments often suffice, even for scarce species. The zones of impoverishment found in groups such as bumblebees are not yet apparent within most fly groups. Examples of fine assemblages surviving in isolated habitat fragments in Warwickshire include 15-plus species of fleshfly (*Sarcophaga* spp.) in abandoned cement quarries and 130 species of hoverfly in partially coniferised ancient woods. These compare favourably with many SSSI's and NNR's that have been well worked for these groups.

The relative success shown by flies in coping with agricultural intensification seems to reflect 4 characteristics:

i) an ability by many to utilise small fragments of habitat.

ii) high mobility and vagility. This is exemplified by the rapid spread of many flies associated with conifers to commercial plantations over much of Britain; also the rapid colonisation of often highly isolated patches of disturbed ground by flies such as tephritids.

iii) the likely ability (which is difficult to measure) of many species to persist in an area at very low population levels. Such species will often respond to certain environmental events with a population boom that suggests they have materialised from nowhere! Having stated this possibility, it is perhaps wise not to be lulled into a false sense of security. Many flies clearly are declining and have requirements that are quantifiable.

iv) the ability of many species to find and utilise temporary habitats such as tall herb and ruderal.

It will be clear from the above that there is much scope for increasing the diversity of flies within agricultural settings, by preserving what established habitat remains and encouraging the creation of features such as wetlands, waterbodies, woodlands where deficient, and by using long term set-aside techniques.

References

Alford, D.V. (1975) *Bumblebees,* Davis-Poynter, London.

Alford, D.V. (1980) *Atlas of the Bumblebees of the British Isles,* Institute of Terrestrial Ecology Publications, No 30, Cambridge.

Chambers, V.H. (1971) Changes in the bee and wasp fauna during the past twenty-five years, *Bedfordshire Naturalist* **26**:52-54

Drake, C.M. (1991) *Provisional atlas of the Larger Brachycera (Diptera) of Britain and Ireland,* Institute of Terrestrial Ecology, Huntingdon.

Dunbar, D. (ed.) (1994) *Saving butterflies - A practical guide to the conservation of butterflies,* Butterfly Conservation, Aylesbury.

Emmet, A.M. & Heath, J. (eds) (1989) *The Moths and Butterflies of Great Britain and Ireland* **7(1),** Harley Books, Colchester.

Falk, S.J. (1991a) A review of the scarce and threatened bees, wasps and ants of Great Britain. *Research and Survey in Nature Conservation* **35,** Nature Conservancy Council, Peterborough.

Falk, S.J. (1991b) A review of the scarce and threatened flies of Great Britain (Part 1). *Research and Survey in Nature conservation* **39,** Nature Conservancy Council, Peterborough.

Fussell, M. & Corbet, S.A. (1992) Flower usage by bumblebees: a basis for forage management. *Journal of applied Ecology* **29:**451-465.

Guichard, K.M. & Yarrow, I.H.H. (1948) The Hymenoptera aculeata of Hampstead Heath and the Surrounding District, 1832-1947. *London Naturalist* **1947:**81-111.

Pollard, E. & Yates, T. (1993) *Monitoring Butterflies for Ecology and Conservation,* Chapman and Hall, London.

Thomas, C.D.; Thomas, J.A. & Warren, M.S. (1992) Distributions of occupied and vacant butterfly habitats in fragmented landscapes. *Oecologia* **92:**563-567.

Thomas, J.A. (1984) The Conservation of Butterflies in Temperate Countries: Past Efforts and Lessons for the Future. **In: Vane-Wright, R.I. & Ackery, P.R.(eds).** *The Biology of Butterflies* :333-353, Academic Press, London.

Warren, M.S. (1990) European Butterflies on the Brink. *British Wildlife* **1(4):**185-196.

Williams, P.H. (1982) The distribution and decline of British bumblebees. *Journal of Apicultural Research* **21:**236-245.

Williams, P.H. (1986) Environmental change and the distributions of British bumblebees (Bombus Latr.). *Bee World* **67:**50-61.

Williams, P.H. (1988) Habitat use by bumblebees *(Bombus spp.). Ecological Entomology* **13:**223-237.

Williams, P.H. (1989) Why are there so many species of bumblebees at Dungeness?, *Botanical Journal of the Linnaean Society* **101:**31-44.

Quantifying effects of landscape connectivity and permeability on farmland.

G.L. Fry
Norwegian Institute of Nature Research, PO Box 5064,
The Agricultural University of Norway, N-1432 Aas, Norway.

Abstract

Most animals surviving on farmland live in patchy habitats. These habitats have become fewer and more isolated in post-war Europe. The situation is currently changing for some types of agricultural landscapes with the introduction of new schemes that aim to increase wildlife habitat on farmland. To be effective, these measures should take account of the landscape requirements of species, e.g., by increasing the degree of connectivity between habitats. This paper proposes a method for quantifying the effect of altering landscape permeability on species dispersal, and hence habitat isolation. Reaction-diffusion models are used to quantify the spatial responses of animals to changes in landscape structure. The need for more empirical data on species dispersal in landscape ecological studies is identified. Case studies of the impact of landscape change on the outcome of butterfly movement models are provided as illustration.

Introduction

One of the basic premises in landscape ecology is that the spatial structure of landscapes influences the population dynamics and long-term survival of plants and animals. Although this is a basic assumption based on the results of wide-ranging modelling and theoretical studies in landscape ecology we often lack empirical data to quantify effects or to verify models. With so many changes underway in agriculture, forestry and other rural land uses we need methods to evaluate their costs and benefits. Such techniques will be useful to predict the impact of restoration schemes, to evaluate conservation measures and to compare alternative development scenarios. One of the landscape metrics often used to describe landscapes is a measure of landscape connectivity. Baudry and Merriam (1985) define *connectivity* as the landscape function which expresses the degree to which sub-populations are interconnected into a functional demographic unit. This contrasts with *connectedness* which refers to the structural links between landscape elements. In practice connectivity has been merely an expression of physical connectedness and has poorly reflected the concept of ecological connectivity. Although many theoretical models have been developed to investigate the negative impact of lowering the degree of connectivity between populations, little information on the factors affecting the landscape-scale movement of species is available. Planners and conservationists often use a broad approach which treats all animals (or taxonomic groups, e.g. birds) as responding in the same way to the structure of landscapes. Alternatively, mapped distances between populations are used to describe their degree of isolation, treating agricultural landscapes as homogeneous. Since we have a poor knowledge of the proximate causes of animal dispersal and its response to landscape structures the reasons for this oversimplified approach are easy to understand, yet may be very misleading. If landscape connectivity is an expression of the degree of contact between sub-populations, then it should be possible to model this even in complex environments. In metapopulation theory the ratio between the probability of habitat patches being colonized by dispersing individuals and patch extinction rates will be highly influenced by the permeability of the various structural components comprising landscapes. Landscape permeability is thus an important component in the spatial structure of populations and to regional population survival (Wu 1993). Although in common language permeability is often used to convey the idea of objects moving across some form of barrier, in most landscape models it is used more precisely to for the processes that affect the rate of movement of animals or plants across a landscape. In this wider context a very favourable habitat, e.g. offering an abundance of food resources, may have low permeability since it negatively affects transition times.

The rapid increase in the number of research programmes and published papers on fragmentation is evidence of the perceived risk to wildlife conservation interests presented by fragmentation. On a broad European scale, habitat loss and modification continues to fragment wildlife habitat and an increasing number of species are classified as vulnerable because of their poor dispersal capability (Hodgson, 1993). In this paper I present a methodological approach to quantify the impact of new developments in terms of landscape-scale dispersal. Results may be used as input parameters to a metapopulation model to examine the potential impact of increasing or decreasing the connectivity between populations. Even in situations where metapopulation models may be inappropriate, the success of introducing new habitats will still be dependent on whether or not they are colonized by immigrants of target species (Thomas 1994).

Methods

Defining connectivity and landscape permeability

Patch connectivity is measured as the probability of an individual reaching from a habitat patch containing species A, to another, being the nearest-neighbour patch. Conceptually, this is the reciprocal of isolation considered in Hanski (1990) and Verboom (1990). Patches experiencing low isolation will have a high probability of receiving individuals from one or more neighbouring patches whereas those with high degree of isolation will have very low or zero values for realistic time spans. For many species movement may be considered as a diffusion process. It is then possible to estimate rates of diffusion from one patch to all others in the landscape across and around landscape elements of different permeability to give a visual and tabular overview of the changes in connectivity caused by a new development. Where animals are territorial, diffusion process may still adequately model the dispersal stage that most animals have during their development. Where animal movements are directional this can easily be built into the model by biasing the direction of flow in the landscape to or away from certain landscape elements such as buildings, roads or important resources. For landscapes with linear features which may act as movement corridors to connect habitat islands this approach retains the concept of connectivity as the reciprocal of isolation rather than some expression of the number of physical contacts between habitat patches. In real landscapes, hedges and other linear habitats may have very high permeability values conducting movement along them, whereas other structures such as roads, streams, or shelter belts may have low permeability values and restrict movement. It is also possible to model the longer transition times caused by an abundance of important resources.

The approach proposed in this paper is;
1. to measure landscape permeability for a range of common landscape structures,
2. to reclassify a habitat map of the study area to produce a landscape permeability map,
3. to incorporate the landscape permeability map as an agent in a reaction-diffusion model,
4. to use inter-patch movement to calculate patch connectivity for the species of interest,
5. to repeat the process for maps of the modified landscape
6. to assess the impact of landscape change on connectivity.
6. to use results as input to (meta)population models to assess the impact of changes.

These landscape changes may be from developments increasing isolation between habitat patches or restoration/conservation measures aimed at increasing connectivity between patches of semi-natural habitat (See Fig 1). There are now available movement data on a wide range of mammals, birds and insects. For the case studies presented here I use the case of a butterfly, the scarce copper (*Heodes viraugreae*), which is declining in Norway. Its normal habitat, open meadows are disappearing and becoming increasingly isolated from each other. Empirical studies of the species have demonstrated that overgrown hedgerows and planted shelter belts can severely restrict its ability to move between fields (Fry and Robson, 1994).

Production of landscape permeability maps

There are several methods whereby habitat maps can be converted to digital map files for analysis in GIS or spatial models. A technique widely used as data capture for spatial analysis is the electronic scanning of aerial photographs or paper maps to form digital databases. These raster format databases are then directly amenable to spatial analysis within many commercially available GIS systems such as GRASS, IDRISI, Map II, SPANS, Arc/GRID etc. (see Green *et al.* 1994). For orthophotos, aerial phots which have had distortions due to optics, tilt, and relief removed, the technique is simple and can be used with inexpensive desktop scanners.

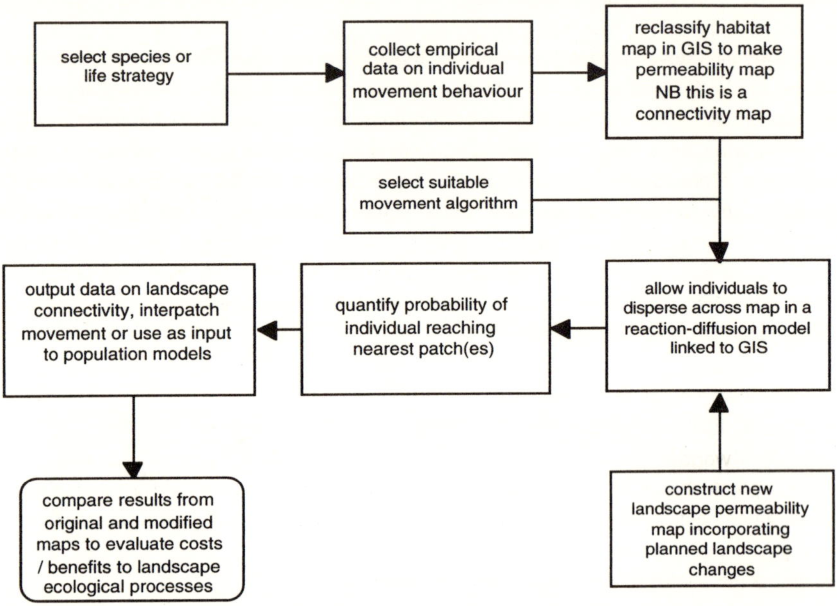

Figure 1. Flow chart showing steps in the use of reaction-diffusion models linked to geographical information systems (GIS) for quantifying the effects of landscape change on landscape connectivity.

Where uncorrected aerial photographs are used, then these may be scanned and corrected by commercially available computer algorithms or used directly to produce digital maps with the aid of an analytical plotter (Warner, 1988). Another way to obtain a landscape permeability map is to purchase available digital maps and add additional features of interest (e.g. from field survey) with the help of a digitizing board. Although many of the available digital maps are available in vector format, these can be quickly converted to raster format within most GIS.

In this study, each pixel on a raster map of the study area was classified by habitat type. A new map was produced by reclassifying habitat variables to permeability values based on the study of Fry and Robson (1994). Each pixel then represented a permeability value to which dispersing animals responded in a modelling environment. When a habitat has a low permeability, say 0.01, then only one in a hundred individuals pass through on approach (routinely this individual is selected by random number generator). At the individual level this is equivalent to only one in a hundred encounters allowing movement to the next pixel. The modelling procedure is described in more detail in Blackburn & Fry (1994).

Scaling

A very important decision at this stage is the selection of map resolution, pixel size, which must be related to the size and movement potential of the species under study. It is inefficient to run models more refined than necessary. For example, a 1m resolution may be too fine to study roe deer, whereas a resolution of 25m will be too coarse for some insect studies. The choice of pixel size will also depend on landscape heterogeneity and the range of permeability values within the study area. In many natural habitats the degree of heterogeneity increases with resolution since many vegetation and soil characteristics can be considered as fractals. This will have the effect of increasing considerably computing time if too fine a resolution is selected whereas too coarse a generalization will fail to reflect the behaviour of individuals in the field.

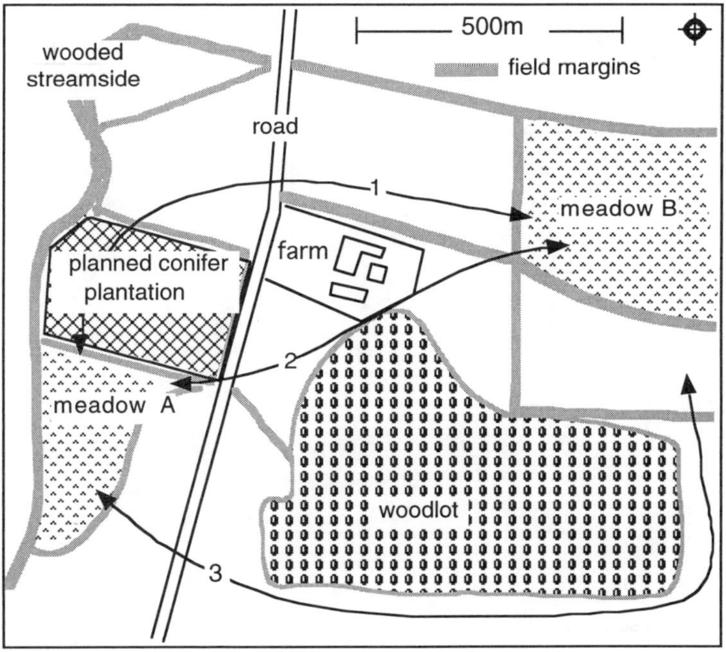

Figure 2a The map shows lowest-cost routes for scarce copper butterflies moving between hay meadows A & B. Map analysis showed route 1 was the optimal route before planting. After planting, route 2 involving crossing field margins, roads and farm yard, and route 3 which circumvents a woodlot became more important but with lower migration probabilities.

The case studies

1. The effects of afforestation on meadow butterflies.

In Fry and Robson (1994) we demonstrated the impact of low hedges and other features on butterfly movement. Results indicated that even low walls and overgrown hedges were significant barriers to butterfly dispersal. More systematic studies with artificial hedges allowed the barrier effect to be quantified. This information was used to classify a farm landscape and to produce a permeability map based on its current structure. A computer simulation of a butterfly population of 1000 individuals of the scarce copper (*Heodes viraugreae*) were released into a meadow and their dispersal across the digitized landscape followed in a reaction-diffusion model (Blackburn & Fry 1994). The landscape was then re-mapped to include proposed forestry plans and the model run again (Figure 2a). Results expressed as probability profiles of butterflies reaching adjacent habitats and the nearest meadow site are shown in Figure2b.

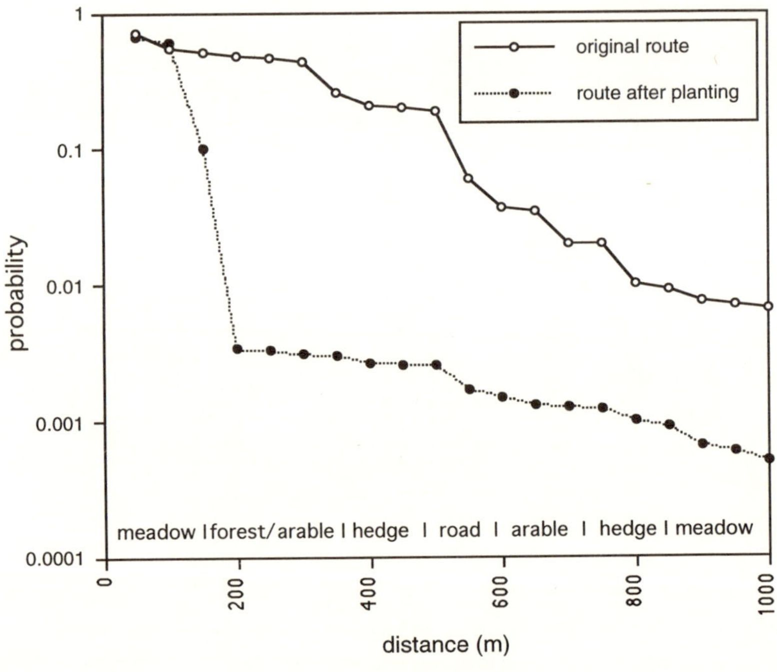

Figure 2b. A profile showing the effect of different landscape elements on the dispersal probability of butterflies at different distances from meadow A to B in Fig 2a.

2. The effect of planting flower seed mixes on the farm-scale distribution of butterflies.

In this simulation the distribution of a butterfly with moderate dispersal ability was allowed to freely disperse along field boundaries reacting to movement barriers and to the distribution of resources. Methods of quantifying flight patterns and general behaviour were based on the work of Dover (1990, 1991), and Banham (1992). The effect of flower distribution on butterfly behaviour was simulated in a reaction diffusion model by setting an attraction field decreasing in density from 0-30m from resources (Blackburn & Fry, 1994). Butterflies landing on resources would stay to feed/rest and then move off to the next flower patch. The process was then repeated with a simulated floral resource planted as a wildflower strip alongside a field margin. The strip was set at 2m wide by 500m long and assumed to be able to produce sufficient nectar not to be depleted. Bees and other potential competitors were not considered. The net result of the processes of attraction, feeding, resting etc. within the study area are shown in Figure 3.

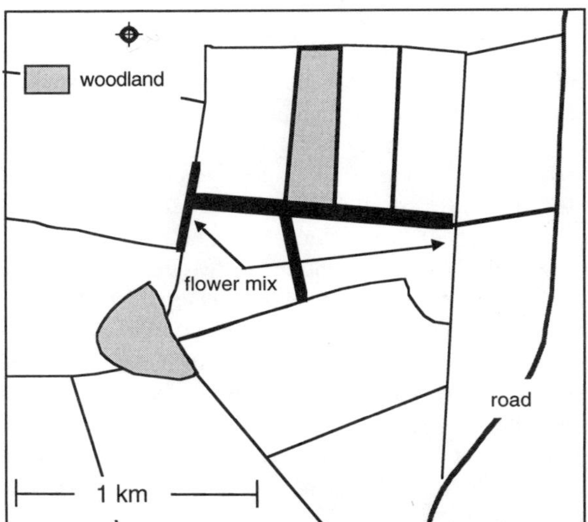

Figure 3. The outcome of a reaction-diffusion model of simulated butterfly distribution 10 days after the flowering of a 2m wildflower mix set alongside a hedge. Line thickness indicates butterfly density (thinnest line represents 1 butterfly per 100m walk, thickest 20 butterflies per 100m). At start butterflies were distributed at random along hedgerows. The manipulated hedge was assigned 1000X the flower resource of the mean value for the others which were allocated at random. The model was conservative and did not include new immigrants, so the result reflect a redistribution of individuals from their random starting points in the landscape.

Discussion

Reaction-diffusion models are developing rapidly and can be constructed to take account of many species-environment interactions simultaneously to quantify their combined impacts on the spatial behaviour of animals. The advent of GIS technology allows spatially explicit environmental and biological factors to be handled as simple data layers or may be spatial models with dynamics of their own. In this way vegetation phenology and other temporal dynamics can be incorporated (Fry, 1994). Population dynamics may be incorporated into models of this type by specifying, e.g., birth, death, immigration and emigration rates which can be programmed to vary in response to other spatially explicit data layers and with time.

In the two simple examples shown here it was possible to demonstrate the potential impact of landscape change on butterfly dispersal. The first case study has been further developed and several alternative planting scenarios compared with model outputs. Impacts can be quantified either through standard cost-distance algorithms which count pixels in each cost class and reflect the connectivity between patches, or by examining the impact of migration on population viability (Wu, 1994).

In the second case I provide an example where altering the permeability of a single element in a landscape, by adding a nectar-rich field margin, can lead to a rapid redispersal of butterflies. Here a low permeability resulted from the low transition times in a highly attractive habitat - a wildflower strip. The same effect has been recorded in the field for conservation headlands in arable farmland (Dover, 1991). The longer-term benefits of such conservation measures will depend on whether they also increase suitable egg-laying sites and larval food plants. The major point from this simple simulation study is that farm-scale abundance patterns can be modelled with very few inputs. The most important need is for empirical data to quantify the way the spatial patterns of resources and landscape structures influence movement behaviour. We must also be aware that low species dispersion can also result from longer retention times in favourable habitats - not usually considered an aspect of landscape permeability. Where resources are especially abundant or attractive, a landscape feature may act as a "sponge" at the farm level - depleting other patches of individuals. By building models to explore the effects of such processes on the farm-scale distribution of individuals we will be better able to interpret field data.

Need for empirical data on spatial behaviour

We have a clear need for empirical evidence quantifying the lifetime dispersal potential of animals. To make a contribution to founding or supplementing small populations these immigrants must reach a new habitat patch while still reproductively active and reproduce there. We can assume for most circumstances this will be directly related to their ability to move between patches of suitable habitat. Future modelling efforts should also incorporate the possibility that only a few percent of individuals in a population may be long distance dispersers. More field data on the influence of sex, reproductive condition and environmental variables on migration are urgently needed.

Incorporation of results into local planning

The rapid development of digital map technology and the use of geographical information systems in planning offices are leading towards the stage where local planning offices are able to compare alternative planning scenarios for a range of impacts. This study concentrates on just one these, connectivity, yet acknowledges the importance of examining other impacts of fragmentation such as the total amount of different habitats lost or created, the quantity and quality of interior habitats with their specialist species, changes in hydrology run-off patterns, and pollution. Modelling techniques clearly offer the possibility to look at the way animals may disperse in real landscapes and can produce dispersal patterns close to those found in field conditions. However, the main value of techniques such as reaction diffusion models is not to predict the outcome of different landscape change but to allow us compare different scenarios. They are not precision tools but can allow the impacts or benefits to be ranked and thus provide an important aid in decision making.

Conclusions

The basis for this method relies on the ability to collect field data on the spatial behaviour of species and from this construct landscape permeability maps. In many cases this will involve detailed behavioural studies backed by experimental work to quantify the effects of a range of landscape structures. For many species this is extremely difficult and the work is time consuming and expensive. However, if the results can be generalized and applied to 'ecological groups' of organisms (Hodgson, 1993), then the work will be a long-term investment of importance for impact assessment. Many modelling initiatives are hindered not by conceptual, programming, or computing limitations - but by the lack of input parameters. Animal movement in response to landscape structure is no exception. We especially need to know what levels of natural variation in movement behaviour to expect and their proximal causes.

Acknowledgements

I thank John Dover for his comments on the project and insights into butterfly behaviour, Wendy Robson for collecting data and keeping my feet on the ground, Nicholas Blackburn for programming and linking animal movement algorithms to digital maps. The work was funded by NINA and the Norwegian Research Council.

References

Banham, A. (1992) *Factors affecting the abundance and distribution of butterflies in abandoned meadows in Southern Norway,* Unpublished BSc thesis, Southampton University.

Baudry, J. & Merriam, H. G. (1988) Connectivity and connectedness: functional versus structural patterns in landscapes. In *Connectivity in Landscape Ecology Proceedings of the 2nd*

International Seminar of the International Association for Landscape Ecology, Münster 23-28.

Blackburn, N.& Fry, G. (1994) Reaction diffusion models with system dynamics at the landscape scale. In *Proceedings of the 2nd CONNECT Landscape Ecology Workshop - Tools for the Analysis of Landscape Processes.,* Danish National Environmental Research Institute (in press).

Dover, J. A. (1991) The Conservation of Insects on Arable Farmland. In N. M. Collins & J. A. Thomas (Eds.), *The Conservation of Insects and their Habitats* Academic Press; London 294-318.

Dover, J. W. (1990). *Butterfly ecology on farmland with special reference to agricultural practices ,* The Cereals & Gamebirds Research Project, The Game Conservancy Trust.

Fry, G. (1994) Landscape ecology of insect movement in arable ecosystems. In D. Glen & H. M. Anderson (Eds.), *Arable Ecosystems for the 21st Century,* John Wiley & Sons Limited, London (in press).

Fry, G. & Robson, W. J. (1994) The effect of field margins on butterfly movement. In N. Boatman (Ed.), *Field Margins: Integrating Agriculture and Conservation,* BCPC Monograph No. 58, Farnham 111-116.

Green, D. R.; Rix, D. & Cadoux-Hudson, J. (Eds.) (1994) *Geographic Information 1994.* Taylor & Francis, London.

Hanski, I. (1994) A practical model of metapopulation dynamics. *Journal of Animal Ecology* **63:** 151-162.

Hodgson, J. G. (1993) Commonness and rarity in British butterflies. *Journal of Applied Ecology* **30:** 407-437.

Thomas, C. D. (1994) Extinction, colonization, and metapopulations: environmental tracking by rare species. *Conservation Biology* *b***8:** 373-378.

Verboom, J.; Schotman, A.; Opdam, P. & Metz, J. A. J. (1991) European nuthatch metapopulations in a fragmented agricultural landscape. *Oikos* **61:** 149-156.

Warner, W. S. (1988) A complete small-format aerial photography system for GIS data entry. *ITT Journal* **1988-2**: 121-129.

Wu, J.; Vankat J. L. & Barlas Y. (1993) Effects of patch connectivity and arrangement on animal metapopulation dynamics: a simulation study. *Ecological Modelling*, **65**: 221-254.

Carabid beetles and landscape structure of agricultural environments - variations at different levels of spatial scale

H. Kinnunen[1] and J. Tiainen[2]

[1] Department of Zoology, P.O.Box 17, FIN-00014 University of Helsinki, Finland
[2] Finnish Game and Fisheries Research Institute, P.O.Box 202, FIN-00151 Helsinki, Finland

---

Abstract

Effect of fragmentation in a Finnish agricultural landscape on carabid beetles (Coleoptera, Carabidae) at local, patch and site levels was studied. At local level total abundance of carabids increased from small isolated to small non-isolated to large patches of farmland, but species number did not vary significantly. At the patch level, both abundance and species number varied significantly between crop types. Proportions of forest and arable land, crop diversity and amount of edge explained patterns of the carabid community in four study landscapes. At the site level, the edge effect was studied. Community composition differed between forest and field, whilst the edge community was intermediate. Few species from the forest penetrated into the arable land and vice versa. Hence the edge function was more a barrier than a buffer. As hierarchy theory proposes, different mechanisms are supposed to affect carabid beetle communities at different spatial scales.

Introduction

Many spatial levels of fragmentedness can be recognised in Finnish agricultural landscapes (see Fig. 1). At the coarsest level, the most obvious feature is the division of land into farmland and other types of land, especially forest. This makes the fragmentation pattern of Finnish agricultural landscapes very different from that of Central and Western Europe where farmland is the prevailing habitat. In Finland, forests cover 55-75% of total area of different provinces, and the proportion of farmland is only 14-23% in southernmost provinces and less in Central and Northern Finland. In a grid of 10x10 kilometres, the proportion of farmland varies from 15 to more than 85% of land area even in the provinces with largest total farmland area, low proportion being most typical (J. Tiainen and J. Sirkiä, unpubl.).

Typically, Finnish farmland occurs in patches which are surrounded by forest, lakes, built-up areas, and roads. The patches can be totally isolated from each other, or they are interconnected by narrow passages with each other. A most conspicuous feature of south Finnish landscapes is the mosaic formed by forest and farmland intermingling with each other in variable relations at many levels of spatial scale. The size of farmland patches varies from less than one hectare to many, sometimes even tens of, square kilometres. The density of, and distances between, patches varies as well.

Finnish fields are small, usually from some tens of ares to a few hectares even in large patches of farmland. This is a consequence of small average farm size, which varies from 14-21 ha of arable field per farm in southernmost provinces (only 5.4% of farms having 50 ha or more; Anon., 1992). Different crops are usually grown on neighbouring fields resulting in high habitat diversity within farmland patches. Fields are usually separated by ditches, farm roads, and wire fences from each other. Often there are even no physical boundaries between two different crops.

Factors affecting animal distribution and abundance in the fragmented Finnish agricultural landscape may manifest themselves differently at different levels of spatial hierarchy. Species may be restricted to certain habitats but they may also depend on higher level dynamics. Theories of island biogeography (MacArthur and Wilson, 1967), metapopulations (Gilpin and Hanski, 1991), and source-sink dynamics (Pulliam, 1988) shed some light on population processes on higher levels of spatial scale. Since different species perceive their environments differently, depending on their movement capacities, one has to pay attention to the levels important to the animal (Wiens, 1989). The populations form a nested hierarchy of spatial inclusiveness beginning with the home range of the individual (grain) and ending with the geographical range of species (Kolasa, 1989). At the level of a grain, factors determining population dynamic processes and, e.g., the pattern of aggregation are certainly other than those determining species distribution at the geographical level.

We are interested in how fragmentation at different levels of spatial scale affect the species number, abundance, and distributions of ground beetles (Coleoptera, Carabidae) in a south Finnish farmland-forest mosaic landscape. The problems we studied for this paper were the following: (1) What is the effect of fragmentedness on a local level, i.e. do carabid communities of isolated and non-isolated small patches of farmland habitat differ from each other and large patches which are supposed to be sources for open-country species? The expectation was that the community is impoverished in terms of species number (S) in small isolated patches compared to small non-isolated patches. Large patches are expected to support most species. (2) What is the effect of fragmentedness on a patch level, i.e. does landscape structure and crop diversity affect the carabid beetle community? (3) What is the effect of fragmentedness at the site level, i.e. how does the field-forest edge affect the beetle community? Does the edge function as a buffer, or is it more like a filter for beetles inhabiting forest and/or open farmland (Forman and Godron, 1992; Fry, 1994)? This paper is a short preliminary report of results from our unpublished research.

Material and methods

The study has been conducted in the surroundings of the Lammi Biological Station of the University of Helsinki, southern Finland (Fig. 1). The experimental designs will be briefly described with results. In all experiments, carabids were collected with pitfall traps (170 ml plastic jars, diameter 70 mm, filled to one third with

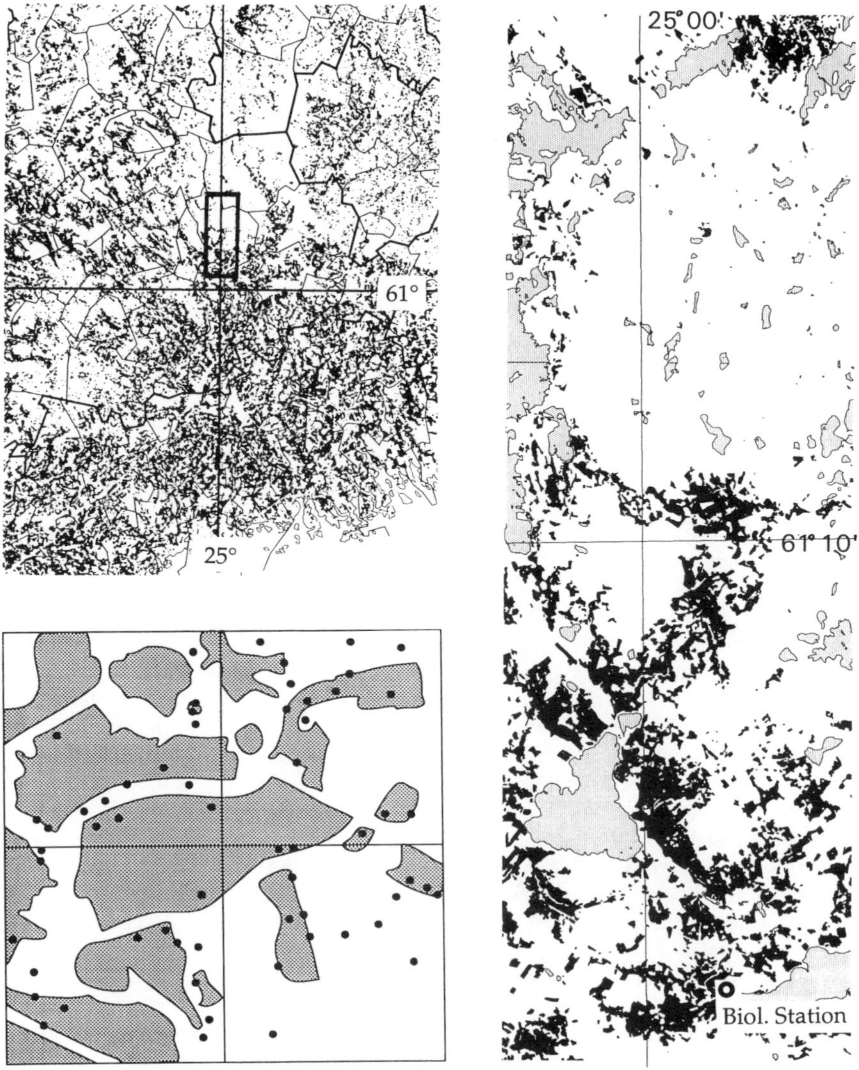

Figure 1. Fragmentation of Finnish farmland in three scales. Dark areas represent arable fields. Upper left: a 150 x 100 km section of South Finland (from Alalammi 1982). Right: a 30 x 10 km section of the study area in Lammi (shown as a rectangular in the upper left panel; lakes shaded; © National Land Survey of Finland). Bottom left: a 1200 x 1200 m study area at the Lammi Biological Station, where the effect of landscape structure on carabid beetle community was studied (dots: sampling sites).

ethylenglycol) during several periods covering the growing season from May/June to late August. The traps were emptied only at the end of a period. Pitfall trapping gives a better measure of activity than true beetle densities (Thiele, 1977), but there is a close relationship between the catch and population size, although this relationship differs among species and within season (Baars, 1979; Ericson, 1979). If trappability of different species varies, their abundance relations in the sample will not reflect the natural ones. However, this bias should be similar in community samples with same species, and does not invalidate a community comparison, if abundance relations of species are not studied within a sample (see Spence and Niemelä, 1994). Our samples to be compared were collected at the same times.

Results

Fragmentation on local level: Effects of farmland patch size and isolation

In order to study the effect of isolation and size of farmland patches on carabids, six small patches both at long (4-7 km) and short (less than 0.5 km from large patches, other small patches in the vicinity) distance from large farmland patches, as well as five large patches were selected. A set of 16 traps was placed in each of the patches for four two-week periods in 1993 (only one period in one of the non-isolated patches). All sites were hay fields.

Both the total number of individuals and species caught varied considerably among the samples (Table 1). The mean number of individuals in a sample was five and almost two times larger in large farmland patches than in small isolated and non-isolated ones, respectively, and three times larger in non-isolated small patches than in isolated ones. The differences were not statistically significant (one-way ANOVA, $F = 2.58$, $df = 2, 14$, $p = 0.11$), however, because of large variation in the data revealed by standard deviations (see Table 1).

Because S strongly depends on the sample size, we rarefied the data before comparisons (see Simberloff, 1978). We excluded one sample from both groups of small patches because of a very small sample size ($N = 26$ and $S = 10$ in one of the isolated patches; and one period in one of the non-isolated patches where $N = 33$ and $S = 13$). The expected number of species in samples of 36 beetles did not differ significantly among the types of farmland patches (one-way ANOVA, $F = 0.27$, $p = 0.77$).

To study the community of farmland patch types which were represented in our samples, we pooled the data and calculated expected numbers of species for small non-isolated and large patches for $N = 400$ and 1200, respectively. The $E(S_{400})$'s did not differ from each other, neither did the $E(S_{1200})$'s in large and small non-isolated patches, respectively (Table 1).

Fragmentation on patch level: Landscape pattern and carabid beetle diversity

Two experiments were conducted to study the effect of landscape structure on carabid communities. In the first one, beetles were collected in sets of nine pitfall

Table 1. Catches of ground beetles in patches of farmland of different sizes and isolation. Total N is total number of beetles caught, S_{obs} the observed number of species, and $E(S_n)$ the expected number of species for a sample of n individuals.

	Small isolated (N = 6)	Small non-isolated (N = 6)	Large (N = 5)
Total N	473	1213	2030
Total S_{obs}	33	42	49
$E(S_{400})^1 \pm SD$	31.7 ± 1.12	31.3 ± 2.13	32.1 ± 2.47
$E(S_{1200})^1 \pm SD$	-	41.9 ± 0.11	44.0 ± 1.84
Mean N ± SD	74 ± 47.3	236 ± 193[2]	406 ± 386
Mean $E(S_{36}) \pm SD$	9.2 ± 2.51[2]	8.6 ± 5.05[2]	10.2 ± 1.78

[1] Calculated from pooled data
[2] Calculated for five patches only

traps, in seven crop types, in six large farmland patches, during six 6-day periods from mid-May to early September in 1987-89 (Tiainen and Tukia, unpubl.).

There were significant differences both in total abundance among the crop types (one-way ANOVA, $F = 2.59$, $df = 6,79$, $p = 0.02$) and number of species (Table 2). The activity density was high in winter and spring cereals and in potato fields, low in sugar beet fields and pastures, and moderate in leys and meadows. The $E(S_n)$ was significantly ($p<0.05$; comparison made by using 2xSD as 95% confidence limits) lower in spring-tilled fields (i.e. spring cereals, potato, and sugar beet) than in fields tilled in the previous year or earlier (i.e. winter cereal, pasture, ley or meadow).

Table 2. Total activity densities of carabid beetles in different crop types in Lammi in 1987-89 (Tiainen and Tukia, unpubl.). Mean is the mean number of individuals per unit effort caught in pitfall traps during several sampling periods, and $E(S_n)$ the expected number of species in a sample of n.

Crop type	Mean ± SD	No. samples	$E(S_{80}) \pm SD$	$E(S_{180}) \pm SD$	Total n
Spring cereal	5.20 ± 4.43	24	14.9 ± 1.79	19.4 ± 1.83	825
Winter cereal	5.62 ± 3.86	3	18.7 ± 1.56	-	141
Sugar beet	1.85 ± 1.63	12	12.4 ± 0.73	-	89
Potato	5.24 ± 4.03	10	13.8 ± 1.79	19.9 ± 0.95	208
Pasture	1.78 ± 1.32	12	20.0 ± 1.53	-	135
Ley	2.65 ± 1.77	11	19.3 ± 1.92	27.5 ± 0.66	187
Meadow	3.87 ± 4.33	14	18.3 ± 2.18	26.1 ± 2.06	374

In the other experiment the impact of landscape structure on the carabid assemblage was studied in a 1200 x 1200 m mosaic of farmland and forest (Kinnunen, 1993). The study area was subdivided into four units (600 x 600 m). The plot was so defined that the subunits represented a gradient where proportion of farmland increased from 30 to 80%. In each subunit five sampling sites were randomly chosen in forests, on fields and at forest-field edges, respectively, 15 altogether in a subunit and 60 in the whole study area. Carabids were caught with sets of nine pitfall traps during four 10-day periods in 1992.

The landscape of each subunit was classified by defining the land cover types of squares in a 60x60 m grid established on the study area. The land use types were potato, spring cereal and sugar beet field, pasture, meadow, human settlement (garden, park), and deciduous and coniferous forest. These data were used to quantify fragmentedness of the landscape by using three indices (diversity, dominance and contagion) to capture its essential characteristics (Turner, 1990). A fourth quantifying index was the length of edge between different land use types. The results indicated that the more uniform the habitats are and the more edges between different kinds of land use types and crops there are, the higher S is. Small plots of a new type of habitat may bring new species to the area, but only larger plots of uniform habitat host a diverse carabid community in the scale studied (36 ha subunits of the study area).

Fragmentation on site level: Edge effect

The edge effect was studied at the field - forest border. We collected beetles from three pairs of 130 m transect lines of 25 pitfall traps. The traps were placed from forest to field every five metres. The furthest traps were hence 45 and 75 metres from

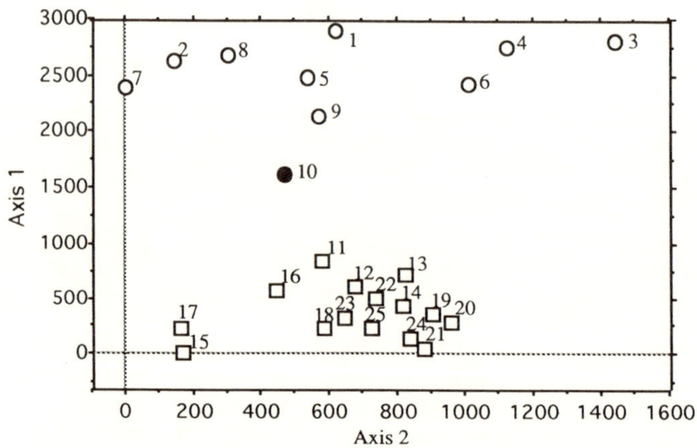

Figure 2. Detrended correspondence analysis on carabid community data from five pairs of pitfall trap lines (traps at 5-m intervals) crossing forest - field edges. Traps from 1 - 9 in forest, 10 on the edge, from 11 - 25 on field.

the edge in the forest and field, respectively. Trap number 10 was exactly on the border of the habitats. Here we were interested in community composition to see whether there are special edge communities and how far from the edge does the edge affect the community.

After pooling, the data from 25 samples were ordinated with a detrended correspondence analysis (DCA) (Hill, 1979). On the first DCA axis, forest and field communities were clearly different. The carabid assemblage in trap 10 was intermediate, and the positions of assemblages next to the edge (traps 9 and 11) were closest to it in their respective groups (Fig. 2). Most of the 42 species collected were found only in either forest or field. Two of the species, *Trechus secalis* (Payk.) and *Pterostichus melanarius* (Ill.), were found in both habitats. Of field species, some few, like *Pterostichus strenuus* (Pz.) and *Patrobus atrorufus* (Ström), were found in forest traps in low numbers, but all species abundant in forests were confined to there. Hence the edge seemed function as a clear buffer between farmland and forest communities. Among individual species there were only few forest species which penetrated the open habitat, and no farmland species were recorded in the forest. The edge assemblage was composed of species of both habitats.

Discussion and conclusions

Hierarchy theories suggest that pattern and process in populations and communities are likely to vary at different levels of spatial scale (e.g. O'Neill *et al.*, 1986). Our results show that at the local level, size of the farmland patch did not significantly affect the species number, but there was variation in the total abundance of carabids (less in small than large patches, less in isolated than non-isolated small patches). At patch level crop type affected both species number and total abundance (less species in spring-tilled crops than in other types of crops, less beetles in more permanent crops than in spring or autumn tilled crops excepting sugar beet). Species number was also affected by landscape structure at patch level (increasing fragmentedness with forest and different crop types decreases, but increasing amount of edges increased the species number). Field - forest edge functioned as a barrier for most of the species, especially for forest species (cf. Halme and Niemelä, 1993).

On the basis of our findings, the pattern of fragmentedness and landscape structure at patch level seem largely determine the composition of, and total abundance in, carabid communities in S Finnish farmland. Although the lower total abundance in small patches compared to large ones suggests that success of carabid populations is affected by patch size, they seem not be at the risk of local extinctions as S was not affected respectively. Site factors are also important. Our results suggest that the effect of the forest-field edge is not important, as edges seemed function as a barrier for most species, but there may be other important site factors.

Acknowledgements

We are grateful to Katariina Järveläinen, Jan Lindström, Harri Tukia and Seppo Varjonen for their help in field, and Katariina Järveläinen, Jari Niemelä and Harri

Tukia for valuable comments on the manuscript. This research has been supported by the Academy of Finland, Kone Foundation, Ministry of Environment, and University of Helsinki.

References

Alalammi, P. (Ed.) (1982) *Atlas of Finland. Folio 231. Farming.* National Board of Survey and Geographical Society of Finland, Helsinki.

Anon. (1992) Yearbook of farm statistics. *Official Statistics of Finland. Agriculture and forestry 1992:2.* National Board of Agriculture, Helsinki.

Baars, M. A. (1979) Catches in pitfall traps in relation to mean densities of carabid beetles. *Oecologia (Berlin)* 41:25-46.

Ericson, D. (1979) The interpretation of pitfall catches of Pterostichus cupreus and Pt. melanarius (Coleoptera, Carabidae) in cereal fields. *Pedobiologia* 19:320-328.

Forman, R. T. T. & Godron, P. N. (1992) Theoretical foundations for understanding boundaries. A. J. Hansen & F. di Castri (Eds). *Landscape boundaries. Consequences for biotic diversity and ecological flows,* Springer-Verlag, New York 236-258.

Fry, G. L. A. (1994) The role of field margins in the landscape. N. D. Boatman (Ed.) *British Crop Protection Council Monograph No. 58: Field margins: integrating agriculture and conservation* 31-40.

Gilpin, M. & Hanski, I. (Eds) (1991) *Metapopulation dynamics: Empirical and theoretical investigations.* Academic Press, London.

Halme, E. & Niemelä, J. (1993) Carabid beetles in fragments of coniferous forest. *Annales Zoologici Fennici* **30**:17-30.

Hill, M. O. 1979 *DECORANA - a FORTRAN program for detrended correspondence analysis and reciprocal averaging.* Cornell University, Ithaca, New York.

Kinnunen, H. 1993 *Maiseman rakenne ja maakiitäjäisyhteisöjen lajikoostumus.* Unpubl. M.Sc. thesis, Department of Zoology, University of Helsinki.

Kolasa, J. (1989) Ecological systems in hierarchical perspective: breaks on community structure and other consequences. *Ecology* **70**:36-47.

MacArthur, R. H. & Wilson, E. O. (1967) *Island biogeography.* Princeton University Press, Princeton.

O'Neill, R. V.; DeAngelis, D. L.; Waide, J. B. & Allen, T. F. H. (1986) *A hierarchical concept of ecosystems.* Princeton University Press, Princeton.

Pulliam, H. R. (1988) Sources, sinks, and population regulation. *American Naturalist* **132**:652-661.

Simberloff, D. (1978) Use of rarefaction and related methods in ecology. Dickson, K. L., Cairns, Jr., J. & Livingston, R. J. (Eds). *Biological data in water pollution assessment: quantitative and statistical analyses,* American Society for Testing and Materials STP 652, Philadelphia, 150-165.

Spence, J. & Niemelä, J. (1994) Sampling carabid assemblages with pitfall traps: the madness and the method. *Canadian Entomologist* 126:(in press).

Thiele, H.-U. (1977) *Carabid beetles in their environments.* Springer-Verlag, Berlin.

Turner, M. G. (1990) Spatial and temporal analysis of landscape pattern. *Landscape Ecology* **4**:21-30.

Wiens, J. A. (1989) Spatial scaling in ecology. *Functional Ecology* **3**:385-397.

The distribution and genetic structure of farmland moth communities

I.P. Woiwod and I.R. Wynne

Farmland Ecology Group, Entomology and Nematology Department, Rothamsted Experimental Station, Harpenden, Hertfordshire AL5 2JQ, UK

Abstract

Long-term data on moth populations from the Rothamsted estate suggest that the fragmentation of the agricultural landscape over the last 50 years has probably had a large deleterious effect on the diversity and abundance of the farmland moth fauna. Recent detailed studies have emphasised the importance of land use in determining the current pattern. The spatial distribution of genetic variation in one species indicates population substructuring associated with habitat fragmentation at the farm scale. The application and utility of genetic markers in estimating levels of gene flow is discussed.

Introduction

Many insect species are declining in industrialised countries. The pressures of modern agricultural methods and, to some extent, increased urbanization has lead to a decrease in the total area of many habitats (through destruction or changes in management regimes) and what remains is often fragmented. This fragmentation may lead to isolation of the populations resident within these fragments, particularly if a species has poor powers of dispersal. Those species which are prone to regular or intermittent local extinctions and rely on recolonisation to persist within a region (i.e. have a metapopulation structure) will decline if habitat fragmentation prevents dispersal between patches. Indeed this appears to be the pattern of decline in many well studied British butterfly species (e.g. Thomas *et al.*, 1992). If conservation management policies are to succeed in preventing further losses in the modern agricultural landscape, the factors influencing the distribution of species and their population structure need to be addressed.

In recent years invertebrate conservation has received a higher profile than in the past. This is partly because of their intrinsic value as major components of biodiversity, but also because of their rapid response to environmental change and thus their importance as indicators of habitat quality (Woiwod & Thomas, 1993). In this paper we describe an attempt to determine some of the factors influencing the distribution of macrolepidoptera, the larger moths, within a modern farm landscape and discuss the use of genetic markers to determine population structure and dispersal ability of insects occupying fragmented habitats.

Long-term effect of farmland fragmentation

Fragmentation might be expected to lead to reduced population size, species loss and genetic isolation. For moth populations on farmland one of the few pieces of quantitative evidence for the long-term importance of the first two of these effects come from the Barnfield light trap on the Rothamsted estate. This trap ran for a four year period from 1933 and again for four years from 1946. In 1960 the trap was reinstated on the same site and has run continuously since.

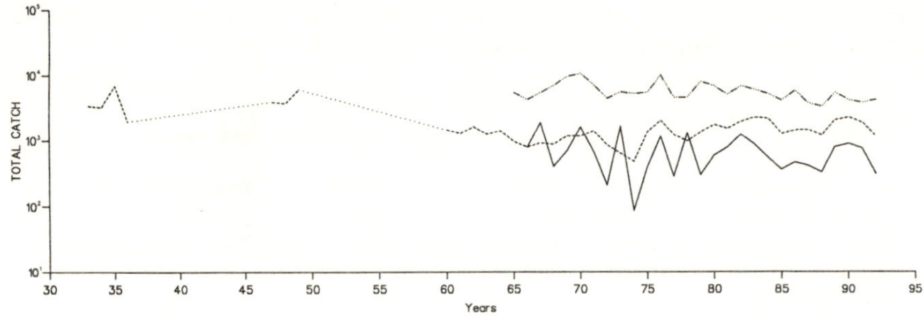

Figure 1. Moth annual totals for the three long-running light traps on the Rothamsted estate. Note the decrease in moth populations in the 1950s at the Barnfield trap (- - -); the high stable population in Geescroft (—··—) and the low fluctuating population at the Allotment site (——). The equivalent diversity plot is given in Woiwod (1991).

The largest change in the moth populations at this site took place between 1950 and 1960 with both the diversity and the total number of the larger moths showing a big reduction (Fig. 1). This period coincided with the mechanisation, intensification and specialisation that took place in agricultural lowland Britain after the war when food was still in short supply. The specialisation was away from mixed farms to arable in the drier eastern counties and dairy farming in the wetter western parts of Britain. Mechanisation led to field enlargement resulting in widespread hedge removal to allow more efficient cultivation. The 1950s also saw the introduction of insecticides and herbicides and so allowed the intensification of agriculture using high inputs of fertilisers and pesticides.

Fragmentation of the farmland landscape for moths is an obvious consequence of hedgerow removal and the resulting loss in connectivity, particularly for the shrub and tree feeding species, but perhaps an equally important but more subtle effect comes from the routine use of herbicides that is part of modern arable farming. This results in the loss of many moth foodplants from the cultivated areas of fields, leaving large areas of a farm inhospitable for the many moth species which feed on low growing herbs. Unfortunately no monitoring of farmland insect populations was taking place at Rothamsted or anywhere else in Britain when these important changes were taking place most rapidly in the 1950s so it is difficult to be certain of the relative importance of these factors in the resulting decline of the moth population on the Rothamsted estate or elsewhere. However, as we show later, land use changes are certainly a very significant factor although the long-term effect of herbicides or insecticides is less clear.

The annual catch from Barnfield is not significantly different from the average of the other arable sites in the Rothamsted Insect Survey's national network so it is probable that similar reductions in moth populations have occurred very widely in the intensive arable areas of Britain (Woiwod, 1991). In recent years with the introduction of set-aside and various environmental schemes there has been a change of attitude in agriculture subsidy away from crop production. Partly as a result many hedges are being reinstated and field margins managed more sympathetically (Boatman, 1994). It is going to be interesting to study the effects of such changes through the moth fauna to see if deleterious changes so far observed can be reversed.

Farm scale pattern of moth distribution

Apart from Barnfield two other sites have been sampled on the Rothamsted estate for long periods of time. One of these is in a piece of mature secondary woodland , Geescroft Wilderness, which has had a light trap since 1965. The other site, sampled since 1966, was originally mature allotments which were due to be destroyed. This site has been much disturbed since and is currently between a building and an arable field. More detailed descriptions of these sites are given in Taylor *et al.*, (1978). The three sites are all within 1000 m of each other and it soon became apparent that there is a considerable amount of spatial heterogeneity in the moth fauna at the farm scale.

In 1990 we were able to extend the sample network on the Rothamsted estate to 26 traps as part of a Joint Agriculture and Environment Programme (JAEP) project. This is enabling us to study the pattern of moth distributions in more detail over the farm and in conjunction with the national light trap network of the Rothamsted Insect Survey (Woiwod & Harrington, 1994) we can begin to look at spatial pattern in relation to scale effects.

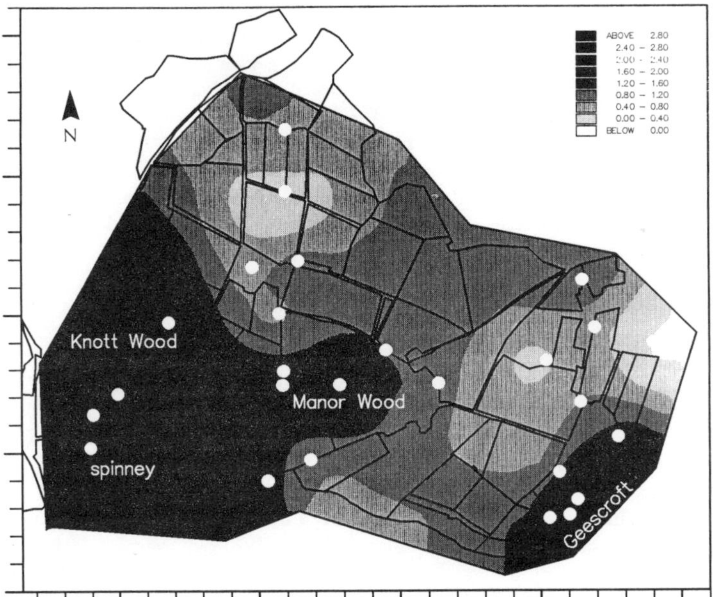

Figure 2. Density distribution of *Epirrita dilutata* in 1990 on the Rothamsted estate ($\log_{10}$ scale). White circles represent light trap sites, tick marks on axis are at 100m intervals.

Each species has its own distinctive distribution of abundance on the farm. Often this is the simple result of foodplant abundance but for many polyphagous and grass feeding species the cause of the pattern is still unknown. Figure 2 shows an example of a moth density distribution, the geometrid *Epirrita dilutata* ([D.&S.]), the november moth. This is a tree feeding species, particularly abundant in the south west of the farm, which has more woodland and mature hedgerows, but which also exists in some numbers in Geescroft Wilderness, a small isolated patch of woodland in the south eastern corner of the estate.

Now that data is becoming available at different scales they can be analysed formally to study levels of aggregation and patterns of spatial variability to see if they have the same structure at different spatial scales, i.e. are fractal, and also assess the consistency of patterns for a particular species through time. So far, there has only been a preliminary analysis for one species, the noctuid *Xestia c-nigrum* (Linn.), the setaceous hebrew character, (Perry, 1994). This is a common species on the Rothamsted estate and in southern Britain generally; the larvae are cutworms and polyphagous on low-growing plants, although seldom reaching pest status in crops. The relationship between the mean and variance of the samples were compared for the years 1990 and 1991 using Taylor's power-law regression. The mean and variance values for both the farmland and south-eastern Britain data lie on a line with a slope almost identical with the published slope for the whole of Britain, suggesting similar patterns of variability of the moth population at different scales (Perry, 1994).

In another analysis of the same species a new class of tests for assessing spatial pattern which takes into account the locations of the spatial units, SADIE (Spatial Analysis by Distance IndicEs), were used to analyse the spatial aggregation of the data (Perry, 1994). The use of these tests provided interesting additional information on the difference in aggregation between years and its interaction with scale. In 1990, a year close to the geometric mean annual catch for this species, counts at both regional (south-east) and farm scale showed strong spatial aggregation. In contrast, in 1991, when densities were about three times larger than 1990 in the south-east and four times larger on Rothamsted farm, spatial aggregation was non-significant at both scales although local heterogeneity was greater relative to the regional heterogeneity than in 1990 (Perry, 1994). Work is in progress to repeat these analyses for further years and for other species which will provide a much better understanding of farm scale spatial dynamics for moth populations.

Moth biodiversity and biomass

Although both moth diversity and biomass have decreased considerably at the Barnfield site since recording started in the 1930s, samples from Geescroft Wilderness suggested that there are still areas on the Rothamsted estate that retain large moth populations with diversities similar to those originally found in open field situations (Fig. 1 & Woiwod, 1991). With the current network of 26 light traps it has been possible to look at patterns across Rothamsted's 330 hectare farm in more detail and relate these to land use in the immediate vicinity of the sample sites. This has been done using a multivariate approach on the 1990 and 1991 data (Woiwod & Thomas, 1993).

The order of importance of land use categories in promoting moth diversity and abundance is much as expected with hedges and bushes adding most to diversity whereas woodland has the greatest effect in increasing moth abundance. Cultivated arable fields and improved grassland are relatively poor but amenity grassland has a particularly impoverished moth fauna (Table 1). These results confirm a previous analysis using data from the national light trap network except in that case only 20 percent of the variability was accounted for because of greater heterogeneity in climate and other environmental variables at the larger scale (Woiwod & Thomas, 1993).

Table 1. Ranking of importance of local land use on diversity and total abundance of the farmland moth community on the Rothamsted estate in 1990.

Land Use (measured as area within 100m radius of trap)	Importance Ranking	
	Diversity (log-series α)	Log_{10}(annual moth abundance)
Bushes and Hedges	1	3
Mixed Woodland	2	1
Deciduous Woodland	3	2
Gardens	4	5
Arable	5	6
Permanent Grass (grazed/cut)	6	4
Unproductive (roads etc.)	7	8
Permanent Grass (regularly mown)	8	7
Variance Accounted for	58%	50%

Genetic structure of farmland moths

Genetic effects of habitat fragmentation

The distribution of genetic variation among populations of a species results from a balance of the forces of mutation, genetic drift, selection and gene flow. Mutation, genetic drift and adaption to local environmental conditions will tend to produce differentiation, whilst gene flow and stabilising selection will tend to produce genetic homogeneity. Genetic drift is the unpredictable change in gene frequency due to finite population size and as population sizes become smaller this effect becomes more prominent. Reduction and fragmentation of once continuous habitat leads to smaller population sizes in the remaining patches and may prevent or reduce gene flow between them. In the absence of gene flow and frequency stabilizing selection, isolated populations will tend to differentiate at a rate determined by the size of the population. Eventually this may lead to the loss of alleles from individual populations and separated populations may become 'fixed' for alternative alleles. In the absence of immigrants natural selection will also tend to favour adaption to local conditions. Whilst in the short term this may increase fitness, in the long term it may reduce the genetic flexibility of the population, leaving it susceptible to environmental change or reducing its potential for expansion. For instance, there is evidence that complete isolation in some British butterfly populations has led to a reduction in dispersal ability (Dempster, 1991). As emigrants rarely encounter or colonise suitable habitats and immigrants never enter the population, selection favours less mobile individuals. However, for many species, it is evident that habitat fragmentation presents little or no barrier to dispersal or (re)colonisation among patches (Hanski *et al.*, 1994).

Measuring gene flow between populations

Direct methods such as mark and recapture can be relatively successful in measuring the dispersal ability of some insect species (e.g. Mckechnie *et al.*, 1975; Hanski *et al.*, 1994). For many insects, however, this is very difficult: The populations may be very large, so that a prohibitive number of individuals may need be marked before any are recaptured. The individuals may be short lived or suffer high mortality and once a marked insect is released the chances of its recapture become less the further it moves (Dempster, 1991). Also, these methods reveal nothing about the reproductive outcome of such movements.

Since the discovery of large amounts of genetic variability at the protein level in the 1970's, and more recently at the DNA level, population biologists have had wide range of genetic markers suitable for making inferences about population structure and testing population genetic theory. The majority of the genetic variation at the DNA and protein levels is believed to be selectively neutral (Kimura, 1968) and thus suitable for investigating population structure and the relative importance of gene flow and genetic drift in natural populations.

At Rothamsted, allozyme studies were conducted on the moths collected by the 26 light traps on the farm. For *E. dilutata* the allele frequencies of three polymorphic enzymes were determined for the woodland localities, where population densities were known to be highest (Fig. 2), from 1990-1992. The three woods in the southwest of the farm had very similar allele frequencies and the differences between them were not significant. Samples from the more isolated Geescroft Wilderness however were significantly different from the other woods (Fig. 3), thus indicating population substructuring even at this local level.

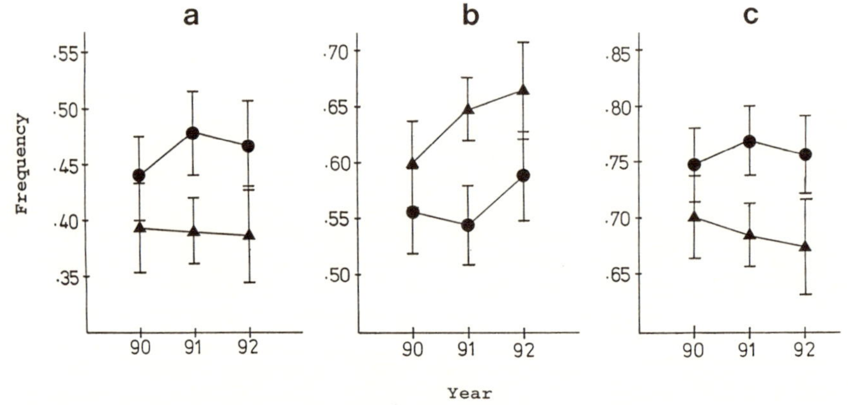

Figure 3. Frequencies of the most common alleles of three enzymes (a: glutamate oxaloacetate transaminase; b: phosphoglucose isomerase; c: phosphoglucomutase) for *E.dilutata* in two woodland localities (Knott Wood ▲ and Geescroft Wilderness ●) separated by 1.5 km. Error bars are 95% confidence intervals.

The level of genetic subdivision among populations can be assessed by measuring the standardized gene frequency variance, F_{st} (Wright, 1951). F_{st} can be used to estimate levels of gene flow between populations, represented in terms of *Nm*, where *N* is the local population size and *m* is the average rate of immigration per generation (*Nm* therefore being the average number of individuals exchanged between populations per generation). This method assumes an island model of population structure (Wright, 1951) and that there is equilibrium between gene flow and genetic drift (Slatkin, 1987). *Nm* can be calculated from the formula $F_{st} = 1/(1+4Nm)$ (Wright, 1951; Slatkin, 1987) if the alleles are selectively neutral. Slatkin (1987) showed that as long as $Nm \geq 1$ significant differentiation between populations would be prevented. Many species presumed or known to be highly mobile have large values of *Nm* estimated from Wright's F_{st} . Estimates of *Nm* from F_{st} values calculated among local populations of *E.dilutata* at Rothamsted fall within this highly mobile species group.

However, estimates of *Nm* based on F_{st} often differ from estimates from independent sources, for example where geographical barriers make the possibility of gene flow extremely remote (Ehrlich & White, 1980; Larson *et al.*, 1984) or when direct observations from mark-release-recapture experiments indicate a different level of gene flow (Mckechnie *et al.*, 1975). These discrepancies are often promoted as evidence that the neutrality hypothesis is incorrect (e.g. Ehrlich & White, 1980) and that instead stabilizing selection is largely responsible for the observed distribution of genetic variation. The neutralists have countered this by arguing that although direct methods may accurately measure gene flow during the period of study, they may miss occasional or periodic events when gene flow is much higher.

However, it is possible that neither selection or recent gene flow events are the major influence on the distribution of alleles and the problem of these discrepancies is due to the assumption (as stated earlier) that the present populations are in equilibrium with respect to gene flow and genetic drift (Larson *et al.*, 1984). A genetically homogeneous population which, due to natural or man made influence, is subsequently fragmented into two or more isolated or semi-isolated subpopulations, may maintain the same genetic structure for some time after its population structure has changed. The period of time required for the new populations to differentiate to the extent where equilibrium is reached (and therefore meeting the requirements of F_{st} to predict *Nm*) will depend largely on the size of the subpopulations and the number of generations per unit time. If the population size is small, differentiation may occur rapidly, but if it is large, a long period may need to have elapsed before equilibrium is reached.

It is therefore important to emphasise the following point when genetic markers are used to assess the mobility of animals of conservation interest. If the populations studied have not had sufficient time since population fragmentation, estimates of gene flow based on F_{st} alone will always overestimate gene flow.

Acknowledgements

We would like to thank Lynda Alderson, Cliff Brookes, Hugh Loxdale, Joe Perry, Adrian Riley, Martin Townsend and members of the Rothamsted Insect Survey for help in various aspects of this work. Parts of this project were funded by NERC under the Joint Agriculture and Environment Programme .

References

Boatman, N.D. (1994) (Ed.) *Field Margins: integrating agriculture and conservation,* BCPC Monograph No 58, BCPC, Farnham.

Dempster, J.P. (1991) Fragmentation, isolation and mobility of insect populations. N.M. Collins & J.A. Thomas (Eds). *The Conservation of Insects and their habitats* Academic Press, London 143-153.

Ehrlich, P.R. & White, R.R. (1980) Colorado checkerspot butterflies: isolation, neutrality and the biospecies. *The American Naturalist* **115:** 328-341.

Hanski, I.; Kuussaari, M. & Nieminen, M. (1994) Metapopulation structure and migration in the butterfly *Melitaea cinxia. Ecology* **75:** 747-762.

Kimura, M. (1968) Evolutionary rate at the molecular level. *Nature* **217:** 624-626.

Larson, A.; Wake, D.B. & Yanev, K.P. (1984) Measuring gene flow among populations having high levels of genetic fragmentation. *Genetics* **106:** 293-308.

Mckechnie, S.W.; Ehrlich, P.R. & White, R.R. (1975) Population genetics of *Euphydryas* butterflies. I. Genetic variation and the neutrality hypothesis. *Genetics* **81:** 571-594.

Perry, J.N. (1994) Spatial aspects of animal and plant distribution in patchy farmland habitats. D. Glen & M. Greaves (Eds). *Ecology and Integrated Arable Farming Systems* Wiley, London (in press) .

Slatkin, M. (1987) Geneflow and the geographic structure of natural populations. *Science* **236:** 787-792.

Taylor, L.R.; French, R.A. & Woiwod, I.P. (1978) The Rothamsted Insect Survey and the urbanization of land in Great Britain. G.W. Frankie & C.S. Koehler (Eds). *Perspectives in Urban Entomology*, Academic Press, New York 31-65.

Thomas, C.D.; Thomas, J.A. & Warren, M.S. (1992) Distribution of occupied and vacant butterfly habitats in fragmented landscapes. *Oecologia* **92:** 563-567.

Woiwod, I.P (1991) The ecological importance of long-term synoptic monitoring. L.G.Firbank; N.Carter; J.F.Darbyshire & G.R.Potts (Eds). *The Ecology of Temperate Cereal Fields*, Blackwell Scientific Publications, Oxford 275-304.

Woiwod, I.P. & Harrington, R. (1994) Flying in the Face of Change: The Rothamsted Insect Survey. R.A. Leigh & A.E. Johnston (Eds). *Long-term Experiments in Agricultural and Ecological Sciences*, CAB International, Wallingford 321-342.

Woiwod, I.P. & Thomas, J.A. (1993) The ecology of butterflies and moths at the landscape scale. R. Haines-Young (Ed). *Landscape Ecology in Britain*, IALE(UK)/Department of Geography, University of Nottingham Working Paper **21:** 76-92.

Wright, S. (1951) The genetical structure of populations. *Annals of Eugenics* **15:** 323-354.

Functional consequences of farm woodland fragmentation on insects, small mammal and bird populations

P. Dennis[1], R. Robertson[2] and S. E. Ferron[2]

[1]Macaulay Land Use Research Institute, Craigiebuckler, Aberdeen. AB9 2QJ UK
[2]Institute of Cell, Animal and Population Biology, University of Edinburgh, Ashworth Laboratories, King's Buildings, West Mains Road, Edinburgh. EH9 3JT UK.

Abstract

The distribution of arboreal arthropods, small mammals and wintering birds were surveyed within a single woodland system in Midlothian, Scotland. Emphasis on variables of habitat composition and extent allowed functional interpretations to be made from results of the pattern analyses. The different fauna exploited the woodland system at different scales but there were common features of the woodland system that limited the distribution of these groups. Phytophage insects were dependent on individual leaves but an exposed canopy or draughty litter layer reduced the value of the site. Likewise, these characteristics were correlated with a poor diversity and structure in the field layer plants, factors which restricted the distribution of bird and small mammal species throughout the available woods. At the functional scale, although the resource requirements of these diverse groups varied widely, there was inter-correlation between the essential habitat components. The extent of these different species could be determined directly from the interaction of the fragmentation and composition with the extrinsic environmental influences of the site.

Introduction

Many studies of animal distributions across a system of woods are undertaken separately for particular groups, there are many examples on insects, birds and small mammals (Usher, Field & Bedford 1993; van Dorp & Opdam 1987; Middleton & Merriam 1981). Many of these studies are difficult to compare not least because of the traditional application of Island Biogeography Theory (MacArthur & Wilson 1967). Woodlands were defined as islands without quantifying the composition of the woodland matrix. It is therefore a simple but useful step to study several organisms with different scales of interaction within a single woodland system. We assume that different species interact with the woodland system at different scales and respond to a range of within and between wood variables which define the scale of interaction of a particular organism. We infer the functional constraints on these species of particular woodland factors and assess the compatibility of their requirements at the scale of human activities. An assessment of the nature conservation value of woodlands could be based on the correlated vulnerability of these wildlife populations under different tree and woodland conditions, often affected by human management.

Figure 1. Determinants of animal species richness in woodlands

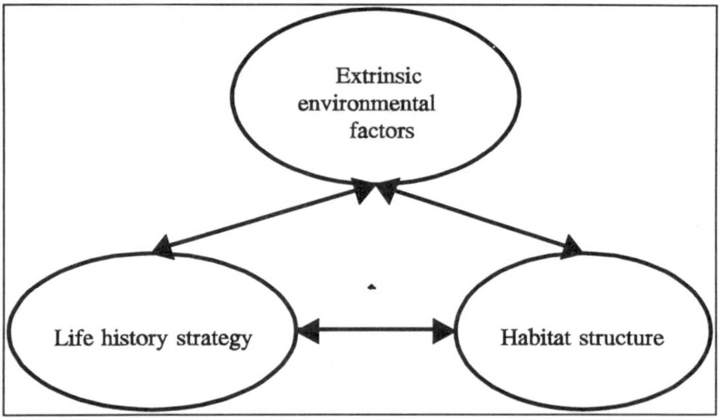

Insect life history, habitat quality and extrinsic environmental influences

Particular tree species can support a greater number of arthropod species in their canopies and beneath in the field layer. This has been determined by the historical abundance, the longevity and structural complexity of the tree species and this apparency through time and space has accrued more species (Warren & Key 1991). The biological diversity of a woodland is therefore enhanced when there are more tree species present and when individual species are represented by a variety of age classes. Natural woodlands are a complex of habitats. Openings in the canopy provide glades which are amongst the richest locations for the invertebrate and floral species. The light penetration combined with the shelter of the surrounding trees promotes nectar providing flowering plants which support the adults of many woodland butterflies (Warren & Key 1991). Factors that limit the distribution of phytophagous species within a woodland system were investigated in Midlothian, Scotland.

A detailed pattern analysis was carried out on phytophagous species on three trees of beech within 15 woodlands. The three beech sampled in each wood were at different locations throughout each wood. Leaf mines and galls were counted on leaves collected by pruning throughout the canopy of each tree. Species richness measurements from leaves collected from the tree canopies were regressed with models composed of within- and between- wood variables. The main factors that limit phytophages on beech were the number of species and depth of field layer, density of leaf litter and connectivity through hedgerows, lines of trees and the density of leaf litter in the wood (Table 1). The species richness of phytophages associated with leaves from particular canopies suggested a causal link with habitat characteristics and extrinsic environmental conditions. Similar variables accounted for the distribution of canopy dwelling spiders and beetles which were sampled by the application of an insecticidal fog to each beech canopy (Table 2).

Table 1. Stepwise variable selection and multiple regression of phytophage species distribution in 1991 against variables that characterised woodland structure and situation in Midlothian, Scotland.

Variable	Coefficient (a, b's & R)	F-ratio (Variable entry and model)	df	P	R^2
INTERCEPT	3.650				
HGTR	0.611	19.24	-	-	-
HFVEG	0.095	5.15	-	-	-
FDIV*					
LITT	0.0001	4.44	-	-	-
MODEL	0.817	27.52	3,41	<0.001	0.64

HGTR, extent of hedgerow or tree network (number x nodes); HFVEG, height of field layer vegetation (cm); FDIV, plant species richness in field layer; and, LITT, mean density of leaves in litter (m^{-2}).

Table 2. Woodland factors which account for the distribution of Araneae and Coleoptera on beech canopies in 1990. Analysis described in Table 1 legend.

Variable	Coefficient (a, b's & R)	F-ratio (Variable entry and model)	df	P	R^2
INTERCEPT	0.880				
DRGHT	-0.040	7.23	-	-	-
HFVEG	0.01	5.55	-	-	-
DEDGE	-0.004	13.52	-	-	-
MODEL	0.720	10.23	3,32	<0.001	0.46

DRGHT, draughtiness of woodland (index); HFVEG, height of field layer vegetation (cm); FDIV, plant species richness in field layer; and, DEDGE, distance to edge of wood (m).

Functional link between wintering passerines and woodland resources

Twenty sample sites spread throughout the wood network were counted six times in February to April 1993. Dual belt point counts of 30 minutes were used to correct for differences in the visibility of species (Bibby, Phillips & Seddon, 1985). The day and time of visits was determined randomly. All counts were made between one to six hours after sunrise at weekly intervals. The effect of a comprehensive range of habitat factors on bird numbers was analyzed in a stepwise regression. Four habitat variables accounted for 75 % of the variance in the species number observed in each wood (Table 3).

Though all the sites are linked into a single network, they supported different compositions of bird species which could be associated with five distinct habitat types. 1. Planted shelterbelts which are derelict and exposed and have little or no herb/shrub cover. 2. Planted shelterbelts in networks that are better sheltered. 3. Large, planted woods dominated by beech, elm and oak. 4. Semi-natural woods with lower canopies that comprise birch and pine species. 5. Younger plantation woods with few beech and many sycamore. The semi-natural woods support a greater number of species than do the shelterbelts and *Erithacus rubecula* (robin) and *Troglodytes troglodytes* (wren) avoid shelterbelts whereas *Parus caeruleus* (blue tit) and *Parus major* (great tit) forage throughout them.

The bird species also used different foraging niches (Colwell & Futuyma, 1971), with the majority of records for each species coming from only two or three parts of the trees or the field layer (Figure 2).

Figure 2. Foraging niches of *Parus major* (GTT), *P. caeruleus* (BTT), *P. ater* (CTT), *Aegithalos caudatus* (LTT) and *Regulus regulus* (GCR). The shaded area shows the commonest location, the hatched area extends this to include 50 % of the foraging records. Species varied by their height above ground (Kruskal Wallis analysis of variance, H = 98.22; P<0.001) and distance from trunk (H = 32.09; P<0.001).

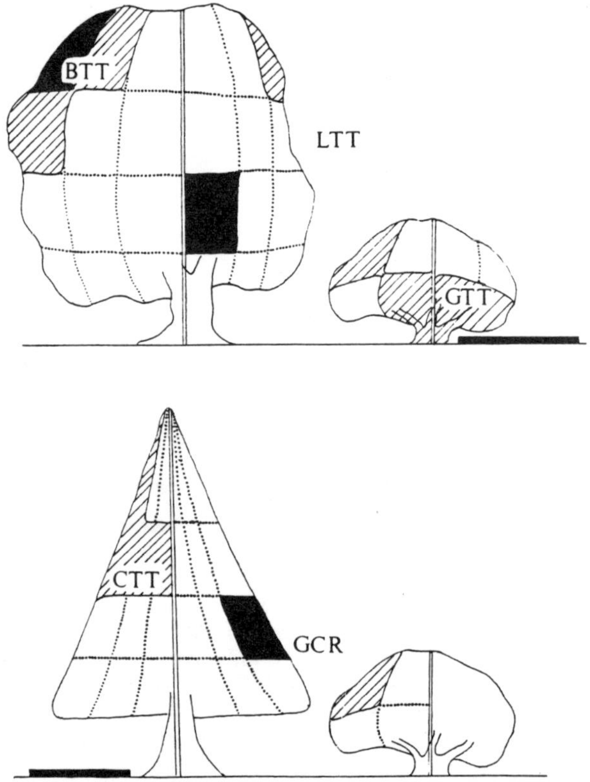

Table 3. Regression of habitat variables against bird species richness (* significant at P<0.05).

Variable	HERB	PEAR	GDIV	CDIV
Correlation coefficient	0.590	-0.616	-0.490	0.372
Variance explained (r^2)	37.3*	14.2*	13.0*	11.1* Total R^2 = 75.6

HERB, percentage field layer dicotyledons; PEAR, perimeter: area ratio for wood; GDIV, diversity of field layer plants; and, CDIV diversity of trees in canopy.

Overall, *P. caeruleus, A. caudatus* and *R. regulus* feed on trees, *P. major* on or close to the ground and *P. ater* primarily on the ground, but occasionally on trees. The broadleaf feeders separate in terms of the heights they occupied whereas the conifer feeders also separated by the distance from the trunk that represented different foraging surfaces.

Small mammal distribution in farm woodlands

Sixteen deciduous semi-natural or farm woods were trapped using Longworth live traps. On each site, forty Longworth traps were placed on a 30 m by 40 m grid. Contrasting sites were paired and trapped for a total of four days. The animals captured were marked by fur clipping to distinguish recaptures. The numbers of each species captured on each site was used as an index of the population density to investigate if there were any differences in the distribution of these small mammals compared with woodland variables. Wood mice, *Apodemus sylvaticus* were captured in all sites and the relative abundance related to the proximity of adjacent woodland cover, the diversity of plant species in the wood and the diversity of the tree canopies (Table 4). These factors could relate to food availability which is an important factor which may determine the distribution of the wood mice in all types of sites.

Table 4. Habitat variables that relate to the abundance of *Apodemus sylvaticus* in a farm woodland system.

Variable	Coefficient (a, b's & R)	F-ratio (Variable entry and model)	df	P	R^2
INTERCEPT	1.870				
GNBLCK	0.010	99.93	-	-	-
CDIV	0.840	19.06	-	-	-
TDIV	-0.150	16.43	-	-	-
MODEL	0.960	40.71	3,13	<0.001	0.90

GNBLCK, shortest distance to next wood (m); CDIV diversity of trees in canopy; and, TDIV, total plant species richness in wood.

Table 5. Habitat variables that relate to the abundance of *Clethrionomys glareolus* in a farm woodland system.

Variable	Coefficients (a, b's & R)	F-ratio (Variable entry and model)	df	P	R^2
INTERCEPT	-5.620				
LGAREA	6.220	24.49	-	-	-
TDIV	0.080	4.78	-	-	-
LITT	-0.030	13.86	-	-	-
MODEL	0.850	8.90	3,13	<0.01	0.65

LGAREA, log area of wood (ha); TDIV, total plant species richness in wood; and, LITT, mean density of leaves in litter (m^{-2}).

Bank voles, *Clethrionomys glareolus* were trapped predominantly in the semi-natural or large planted woods. The size of the wood was a powerful estimator of their relative densities (Table 5). Relative density of the voles also related to plant species richness and litter density. These data suggest they are supported by woodlands with a structure typical of natural woodland, such as species and age diversity, and stratification of the vegetation, features absent from small, remnant and exposed woods. It is however, possible that the larger woods act as a source of bank voles for colonisation of some of the poorer woods in the summer.

Discussion

When attributes of the extent and quality of habitat, extrinsic environmental factors and the life history strategy of the organisms were combined, patterns in the distribution of the organisms could be given a functional interpretation. A high proportion of trees in small, exposed woods are exposed to the weather. The leaves of these trees provide a poor habitat in summer and the leaf litter is depleted in winter to a greater degree than under trees in larger woods or small woods made draught free by edge shrubs and a well structured field layer. The groups studied interacted with an individual woodland system at different scales. Resource availability and use had to coincide at the right scale to determine the success of populations of these different organisms (Figure 3).

Results from Longworth trap based surveys of small mammals showed that their distribution was linked to the distribution of woods of semi-natural structure. For *Apodemus sylvaticus*, a combination of tree cover and structural complexity was required in the field layer, factors that are associated with the structural requirements of the arboreal arthropods. Bird species also required resources available from the full range of structural complexity available within the semi-natural woodlands. Certain features of these woods seem more important in determining species richness of birds. The larger woods which are less draughty, offer a denser herb layer and a well developed shrub layer. In contrast to the dichotomy in the literature, habitat structure or area as the determinant of species richness (Begon, Harper & Townsend, 1990). These results suggest that both are correlated with species richness. As most the woods supported some combination of the commonest edge species (*Fringella coelebs* (chaffinch) and *P. caeruleus*), these habitat features are attracting the more typically woodland species, such as *Certhia familiaris* (treecreeper) and *A. caudatus*.

In general, individual bird species densities do not seem closely correlated with single habitat features. Birds may therefore select sites on the basis of an overall complex of cues (the habitat gestalt; Cody, 1985). Habitat selection may be hierarchical, first at a broad scale to the gestalt then on a finer scale to individual features (eg. Morton, van der Voort & Greenberg, 1993). *P. caeruleus* and *P. ater* forage throughout a large area in winter (Lack, 1986), moving between trees. *P. major* remain in their breeding territory throughout the winter and are dependent on the summer aphid and caterpillar populations within the territory. The numbers of bird species recorded in a wood within the system are not therefore a function of the geographical parameters that characterise a particular wood, extolled in previous studies. The functional approach takes account of the phenology of these species. Records of bird activity by mobile, edge species like *P. caeruleus* and *F. coelebs* or typical woodland species like *C. familiaris* and *A. caudatus* related to the aggregate of resources available across the woodland system. If the required resources are unavailable across all the woods, or deficiencies occur through the season within potential territories, the species do not occur. Similarly, the endophage arthropods may be limited by depletion of the leaf litter in winter and it is possible that the exposure of the leaf canopy to wind in summer caused the similar pattern of distribution of spiders, weevils and predatory beetles, for instance from disruption to web building or search activities.

Figure 3. Functional relationships between wildlife populations (ellipses), woodland structure (hexagons) and weather (rectangles) in a fragmented woodland system.

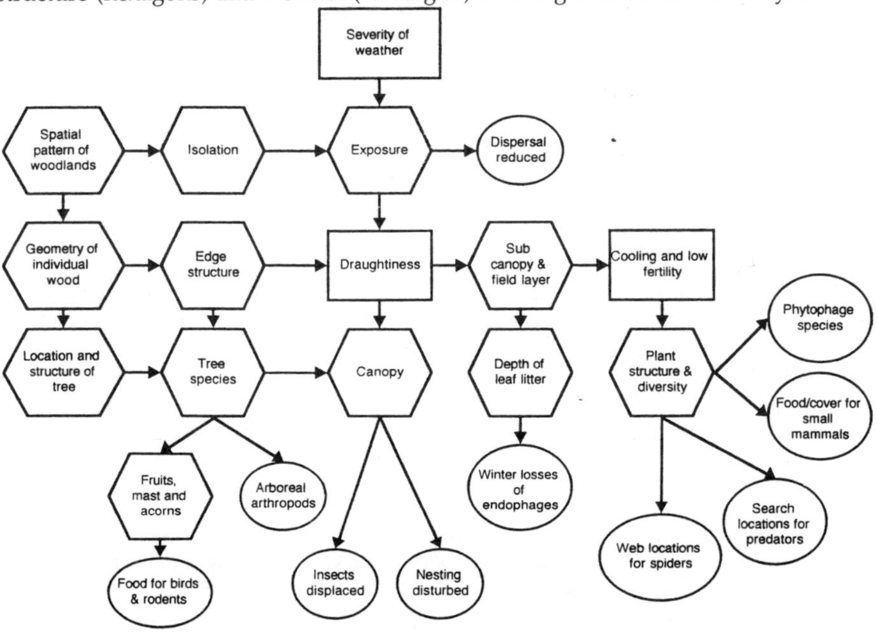

This approach is useful for building a picture of the resources, both similar and unique, that must be present in a woodland to maximise the number of natural species that can be supported. Emphasis on the functional scales of interaction allow more meaningful interpretation of the pattern of individuals amongst habitat patches than the traditional island biogeographical approach that uses a rigid, usually human perceived scale (see Haila, 1990).

Acknowledgements

P. D. was funded by the FRCC Special Topic in Farm Forestry. The work was carried out at the NERC ITE Edinburgh Research Station. P. D. is indebted to Allan Watt for his support and advice.

References

Bibby, C. J.; Phillips, B. N. & Seddon, A. J. E. (1985) Birds of restocked conifer plantations in Wales. *Journal of Applied Ecology* **22:** 619-633.

Begon, M., Harper, J. L. & Townsend, C. R. (1990) *Ecology: individuals, populations and communities*, 2nd edition, Blackwell Scientific Publications, Oxford.

Cody, M. L. (1985) *Habitat selection in birds*, Academic Press, Orlando. pp. 3-56.

Colwell, R. K. & Futuyma, D. J. (1971) On the measurement of niche breadth and overlap. *Ecology* **52:** 567-576.

Haila, Y. (1990) Toward an ecological definition of an island: a northwest European perspective. *Journal of Biogeography* **17:** 561-568.

Lack, P. C. (1986) *The atlas of wintering birds in Britain and Ireland*, T. and A. D. Poyser, Calton, UK.

MacArthur, R.H. and Wilson, E.O. (1967) *The theory of island biogeography*, Princeton University Press, New Jersey.

Middleton, J. D. & Merriam, G. (1981) Woodland mice in a farmland mosaic. *Journal of Applied Ecology* **18:** 703-716.

Morton, E. S., van der Voort, M. & Greenberg, R. (1993) How a warbler chooses its habitat: field support for laboratory experiments. *Animal Behaviour* **46:** 47-53.

Usher, M. B., Field, J. & Bedford, S. (1993) Biogeography and diversity of ground dwelling arthropods in farm woodlands. *Biodiversity Letters* **1:** 54-62.

van Dorp, D. & Opdam, P. (1987) Effects of patch size, isolation, and regional abundance on forest bird communities. *Landscape Ecology* **1(1):** 59-73.

Warren, M.S. & Key, R.S. (1991) Woodlands: past, present and potential for insects. In: *The conservation of insects and their habitats* (N.M. Collins and J.A. Thomas, eds) Academic Press, London. pp. 155-211.

Birds in a mountain landscape

A. Farina

Lunigiana Museum of Natural History and Laboratory of Landscape Ecology, 54011 Aulla, ITALY

Abstract

39 species of birds were found during the 1993 breeding season in a 103 ha of summit fragmented prairies of the Northern Apennines; of these 27 are breeding in the study area. The bird assemblage is strongly conditioned by the ecotone between prairies and the beech forest. Altitude, orientation, vegetation cover type, distance from ridges and from forest edges in different manner are the key factor shaping the structure and the dynamic of this bird assemblage. *Anthus spinoletta* is the only species exclusive of this landscape.

Introduction

The fragmentation of the natural landscape is a common pattern in most of the world regions, but is particularly evident in the Mediterranean due to the pluri-millenarian human disturbance regime.

The fragmentation affects the distribution of plants and animals with consequences on many ecological processes (Saunders *et al.* 1991).

One of the more evident consequences of the fragmentation is the isolation of populations and their organization in meta-populations (Levins 1970, Opdam 1991) separated by a matrix of unsuitable habitats (Forman & Godron 1986) and genetically connected by emigration/immigration movements (Hansson 1991).

The case presented here has, as main objective, an investigation of the structure and dynamics of an assemblage of birds in fragmented mountain prairies on the Northern Apennines. These prairies, that appear as islands of different shape and size, are distributed in a narrow belt, on average, from 1500 to 2000 m, long the mountain ridge and are surrounded by a dense cover of beech forest.

In these terrestrial islands the structure and dynamic of the bird assemblages are influenced by different ecological processes such as isolation, edge effect, climatic slope orientation constraint, etc.

Study area

The study area, 103 ha, is located on the Cavalbianco Mountain (44° 17′ N, 10° 18′ E), an isolated peak , 1850 m in altitude, of the Northern Apennines.

The pyramid-like shape of this mountain allows different microclimatic

conditions according to the slope orientation.
The soil is composed of oligogenic macigno with outcrops distributed especially on the south-eastern slopes.
The investigation was limited to the summit prairies and the beech forest edge.

Fig. 1 Distribution of the mountain prairies in the northern Apennines and location of the study area

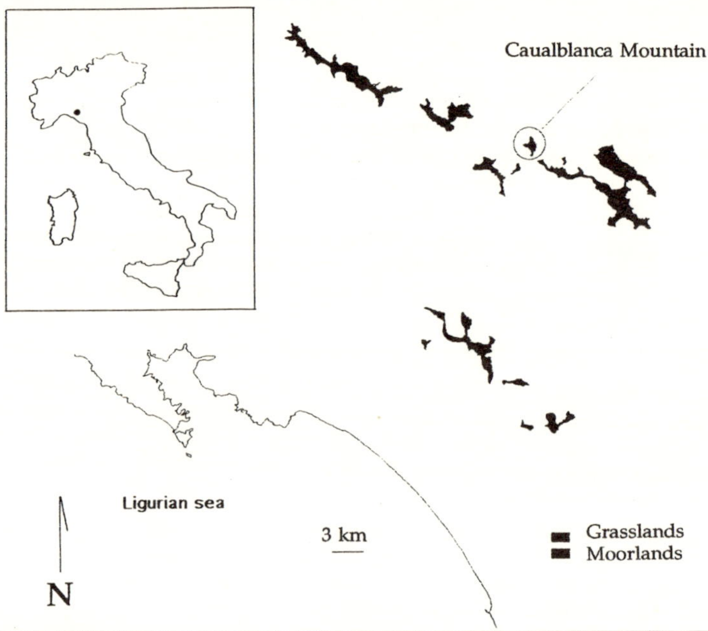

The landscape

The summit prairies are open spaces like islands surrounded at the base by a dense cover of beech forest. The edge between these two elements of the mountain landscape is marked by a linear structure of old trees maintained to protect woodland against snow slides in spring time and to assure shelter to livestock during summer storms.
Until the recent past the summit landscape was characterized by a strong human disturbance regime. This landscape was utilized for livestock grazing and the surrounding beech forest was exploited for charcoal production (Farina 1991).
A core of man-made clearings, in the past connected with summit prairies, is included in the beech forest matrix.
The herbaceous vegetation has been strongly influenced by grazing regime and to day shrubby plants are quickly replacing geophyte and therophyte species typical of the grazed prairies.
The steep slopes are shaped by livestock trails that persist after decades of abandonment.

The vegetation

The vegetation, until the recent past, was strongly disturbed by the grazing regime of livestock (sheep and cattle), but despite this disturbance the vegetation has conserved a very rich assemblage of endemic and alpine relict species.
Four main types of land cover were observed : *Myrtillus* sp.p. moorland (58.56 ha), *Brachypodium genuense* prairie (30.28 ha), light vegetation (5.56 ha), rock outcrops (3.88 ha).
The distribution, variety and structure of vegetation is strictly linked to the orientation and slope steepness.
The south and south-western slopes are dominated by *Brachypodium genuense* in some parts substituted by *Calluna vulgaris*. The northern and eastern slopes are covered by *Vaccinium myrtillus, V. uliginosus, V. vitis idaea.* and *Juniperus nana*. The shrubby species are actually invading most of the grass spaces.

Methods

Birds were censused during 17 sessions from April 28 to August 14, 1993 , the study area was crossed long several transects, 50 m apart. These were chosen to ensure the efficient survey of the area and totalled 62 hours overall with a mean duration of 3.65 hours *per* census session. Birds were counted only if close to the observer, no further than 30 m away. Song, call , alarm call, and other behaviours such as chases, fights, feeding young, etc. were also recorded.
Censuses were carried out in the morning, one hour after sunrise and until 10 - 11 a.m , in good weather conditions, no rain or strong wind,.
Birds were located by using a GPS (GPS Pathfinder Systems, Trimble Navigation®) facility and the bird data entered directly in a data logger (Psion Organizer II LZ64, Psion®). To improve the accuracy of the bird position a differential correction of the field data was obtained by using a GPS Pathfinder Basic Station®.
A GIS with a grid cell of 20x20 m was prepared for the study area. The topography and the vegetation cover detected by an aerial photograph of 1989 were included in the GIS together with bird location. The vegetation cover was classified using the GAIA® software. The GIS was prepared by using the MacGIS software (Larsen & Hulse 1989).
The bird analysis was made at three temporal scales: bi-weekly (7 periods), monthly (4 periods) and bimonthly (2 periods).
Five landscape attributes: altitude, orientation, distance from ridges, distance from tree line and vegetation type cover were utilized to investigate the spatial distribution of birds. The bird diversity was calculated using the Shannon-Wiener index of diversity (Shannon & Weaver 1962) . The Systat® program was used to make the statistical analysis.

Results

39 species were found in the study area (Table 1), of these 27 species were breeding in the study area, the others were breeding in the neighbouring spaces or were visiting the study area for foraging activity. Of these *Turdus pilaris* is a migratory bird that spends the non-reproductive period in the Mediterranean basin and *Alectoris rufa* is probably breeding in the study area but the reduced number of sightings has not confirmed reproductive success.
The richness and the abundance are relatively stable across the different time scales (biweekly, monthly and bimonthly) see Table 1.

Table 1 Number of species and mean abundance at 3 temporal scales

BS	S_{bs}	AB_{bs}	M	S_m	AB_m	BM	S_{bm}	AB_{bm}
1	24	26.32	1	25	23.17	1	28	22.41
2	18	19.18	2	19	21.92	2	25	24.92
3	18	23.45	3	24	22.65			
4	11	17.55	4	16	33.74			
5	17	25.67						
6	22	20.19						
7	16	33.74						

BS: Biweekly intervals; S_{bs}: Biweekly Richness; AB_{bs}: Biweekly mean abundance *per* hour transect; M: Monthly intervals; S_m: Monthly Richness; AB_m: Monthly mean abundance *per* hour transect; BM:Bimonthly intervals; S_{bm}: Bimonthly Richness; AB_{bm}: Bimonthly mean abundance *per* hour transect;

The diversity (Shannon index) of this assemblage is 2.10 and the evenness .57.
Only *Anthus spinoletta* is an exclusive summit breeding bird. *Anthus trivialis, Oenanthe oenanthe, Phoenicurus ochruros* and *Acanthis cannabina* are breeding in mountain range and also at lower altitudes. All the other species are also common in other altitudes and landscapes (Farina 1981, Illner *et al.* 1992). Some species such as *Sylvia atricapilla, Erithacus rubecula, Phylloscopus collybita* are very common in many landscapes and others such as *Alauda arvensis* are common in open grasslands and large fields from the plain to the mountain.
Using to the preferred habitat, it is possible to distinguish five groups of birds: 1. *Anthus spinoletta, Alauda arvensis* (open prairies) , 2. *Phoenicurus ochruros, Oenanthe oenanthe, Monticola saxatilis* (rock outcrops), 3. *Anthus trivialis, Phylloscopus collybita* (edges and clearings) , 4. *Erithacus rubecula, Parus ater, P. caeruleus, P. major, Garrulus glandarius* (forest) , 5. *Acanthis cannabina* (open shrubland).
Anthus spinoletta, Alauda arvensis , Anthus trivialis and *Phylloscopus collybita* were found dominant (abundance ≥5% of the all observations).
The total abundance is highest in *Brachypodium* prairies (250 individuals x 10 ha), followed by rock outcrops (141.75), light vegetation covered soil (127.69) and *Myrtillus* moorland (123.29).

Table 2 Total number of contacts and breeding status during 62 hours of census

Species	TC	BST
Anthus spinoletta	672*	Br
Alauda arvensis	121*	Br
Anthus trivialis	111*	Br
Phylloscopus collybita	73 *	Br
Erithacus rubecula	61	Br
Phoenicurus ochruros	56	Br
Acanthis cannabina	46	Br
Parus caeruleus	33	Br
Oenanthe oenanthe	31	Br
Garrulus glandarius	25	Br
Apus apus	23	Vs
Cuculus canorus	21	Br
Sylvia atricapilla	18	Br
Pyrrhula pyrrhula	17	Br
Falco tinnunculus	14	Vs
Parus major	12	Br
Fringilla coelebs	9	Br
Parus ater	7	Br
Turdus merula	5	Br
Hirundo rustica	5	Vs
Turdus torquatus	4	Br
Alectoris rufa	2	Br?
Monticola saxatilis	2	Br
Accipiter nisus	2	Vs
Turdus viscivorus	2	Br
Turdus philomelos	2	Br
Lanius collurio	1	Vs
Aegithalos caudatus	1	Vs
Sitta europaea	1	Br
Dendrocopus major	1	Br
Picus viridis	1	Vs
Muscicapa striata	1	Br
Prunella collaris	1	Vs
Serinus serinus	1	Vs
Buteo buteo	1	Vs
Certhia brachydactyla	1	Br
Troglodytes troglodytes	1	Br
Lullula arborea	1	Vs
Turdus pilaris	1	Mg

TC:total number of contacts during the census period; BST:breeding status (Br=breeding, Vs=breeding in neighbouring areas, Mg=migrating); *=Dominant species (relative abundance≥5% of the all observations).

The total abundance of birds decreases with the increase of distance from the tree line. Close to this border we have recorded the maximum diversity and abundance. The edge effect is very evident. The density of species has been respectively 180 individuals on 10 ha in the first belt (0-150 m from the tree line), 80 in the second belt (150-300 m), 22.5 in the third belt (300-450m), 25 in the fourth belt (450-600 m).

During the season only *Anthus spinoletta* showed an intra-mosaic shift according to four landscape descriptors: altitude, orientation, ridge and forest edge distances (see fig. 2).

Fig. 2 - Distribution of *Anthus spinoletta* in the prairie landscape according to: altitude; slope orientation; ridge distance; edge distance long the breeding season. In the Z axis the abundance, X axis census session, in Y axis respectively: a. the altitude, b. the orientation (9 categories: 1=North, 2=North-East, 3=East, 4=South-East, 5=South, 6=South-West, 7=West, 8=North-West, 9=flat) , c. the distance from the ridge, d. the distance from the edge.

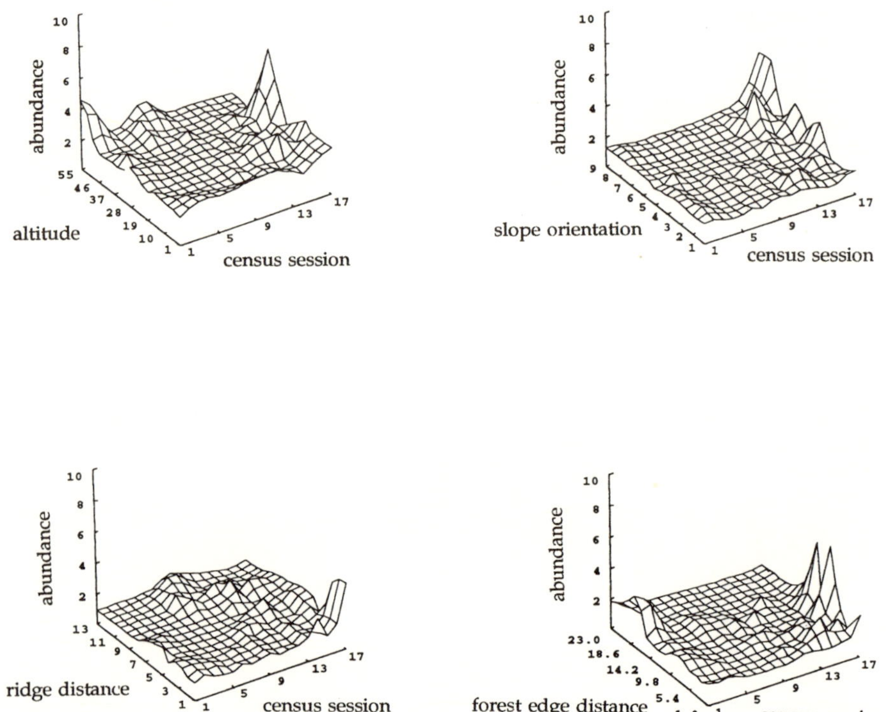

Discussion

The summit prairies are important components of the mountain landscape. The limited size of these prairies creates an ecotonal border that attracts many birds with different habitat requirement. According to the ecotone paradigm (Hansen *et al.* 1988) we can consider these prairies as an extended ecotone and the richness of the bird assemblage an edge effect (Leopold 1933). But many other factors such as altitude, distance from ridges and from forest margin, orientation and vegetation types are constraint factors responsible of bird distribution and abundance. In particular *Anthus spinoletta* appears particularly sensitive to these factors throughout the breeding season.

The intrinsic complexity of this landscape, alternance of shrub patches with outcrops and moorlands allows the presence of a variety of species of different habitat requirements. The mountain character of this landscape is not reflected in the bird assemblage, but is mostly composed of species also living in other areas and altitudes.

The evolution of the vegetation, after the historical grazing disturbance regime, is expected to evolve into a dense shrubby woodland with an expected decrease of bird species richness and abundance.

References

Farina, A. (1981) Contributo alla conoscenza dell'avifauna nidificante nella Lunigiana. *Boll. Mus. St. Nat. Lunigiana* 1:21-70.

Farina, A. (1991) Recent changes of the mosaic patterns in a mountainous landscape (north Italy) and consequences on vertebrate fauna. *Options Mediterranéennes* 15:121-134.

Forman, R.T.T. & Godron, M. (1986) *Landscape ecology,* John Wiley and Sons, New York, USA.

Hansen , A. J. ; di Castri, F. & Naiman, R.J. (1988) Ecotones: what and why? F. di Castri, A.J. Hansen & M.N. Holland (Eds.). *A new look at ecotones.* Biological International, special issue 17:9-46.

Hansson, L. (1991) Dispersal and connectivity in metapopulations. *Biological Journal of Linnean Society* **42**:89-103.

Illner, H.; Judas, M.; Junemann, K. & Oelke, H. (1992) Italian Summer Bird Population (Lunigiana, NW Tuscany, Apennines). *Boll. Mus. St. Nat. Lunigiana,* 8:13-94.

Larsen, K. & Hulse, D.W. (1992) *MacGIS 2.0. A geographic Information System for the Macintosh. The user's guide 2.0.* University of Oregon, Eugene, Oregon.

Leopold, A. (1933) *Game management.* Schriber, New York, USA.

Levins, R. (1970) Extinction. M. Gerstenhaubert (Ed.) *Lectures in mathematics in the life sciences. Some mathematical questions in biology*. American Mathematical Society, Providence, Rhode Island, 77-107.

Opdam, P. (1991) Metapopulation theory and habitat fragmentation: a review of holartic breeding bird studies. *Landscape ecology* **5**:93-106.

Saunders , D.; Hobbs, R.J. & Margules, C.R. (1991) Biological consequences of ecosystem fragmentation: a review. *Conservation Biology* **5**:18-32.

Shannon, C.E. & Weaver, W. (1962) *The mathematical theory of communication*. University of Illinois Press, Urbana, IL.

Persistence, loss and replacement of woodland birds in habitat patches in an arable landscape

S.A. Hinsley, P.E. Bellamy and I. Newton

Institute of Terrestrial Ecology, Monks Wood, Abbots Ripton, Huntingdon, Cambridgeshire, PE17 2LS, UK

Abstract

Turnover of individual adult birds between consecutive breeding seasons was investigated in 16 small woods set in an arable landscape. The majority of marked birds disappeared between breeding seasons. These high turnover rates, coupled with the known good mobility of birds and evidence of local bird movements obtained from marked individuals suggested that, for the particular species studied, the bird populations of small woods behaved more as a single population rather than as metapopulations. Few of the replacement birds present in the second breeding season bred in their natal woods; the majority of replacements were immigrants. These results suggested that small woods might be population sinks, but without knowing the fate of the missing juveniles this conclusion could not be supported. The rate of local extinction of species was substantially higher than would have been recorded without a marked population. Species often persisted in a wood, but all the individuals in the second season were replacements.

Introduction

The fragmentation, and consequent loss, of contiguous habitat is one of the major causes of a worldwide, and accelerating, reduction in biodiversity (Wilson, 1992). In the initial stages of fragmentation, the areas of altered habitat appear as "islands" in a matrix of the original landcover. Later, this arrangement is reversed, with remnant patches of the original continuous habitat becoming the islands in a matrix of the new, imposed land-use. Rolstad (1991) described such a progression from contiguous to fragmented habitat and suggested that the increasing isolation of the fragments could generate metapopulation structures for the inhabitants of the fragments. The degree of isolation required would vary between species according to their dispersal abilities and the nature of the intervening landscape. It has also been suggested that, compared to contiguous habitat, small patches may represent poor quality habitat, in particular in relation to increased levels of predation (e.g. Paton, 1994) and lack of resources (Fahrig & Merriam, 1985). Such patches may therefore act as population sinks (mortality exceeds natality) and thus depend on immigration from source populations (natality exceeds mortality generating a surplus) to avoid extinction.

In this paper, we present results from a study measuring turnover at the level of the individual bird, in populations breeding in small habitat patches. We investigate the possibilities that the bird populations of small woods in an arable matrix have a metapopulation structure (*sensu* Levins, 1970) and that they represent population sinks. These results, mainly concerned with the changes recorded between the breeding seasons of 1990 and 1991, are only preliminary, fieldwork and analysis are both still in progress.

Methods

The study was carried out in 16 small woods (mean area = 0.51 ± 0.27 ha) in Cambridgeshire and south Lincolnshire. The landscape surrounding the woods was largely devoted to intensive arable agriculture, the main crops being cereals, oilseed rape, sugar beet, potatoes and field beans. The loss and replacement between breeding seasons of individuals (breeding adults) of certain bird species was assessed using a marking program. Birds were caught using mist nets and marked individually with a numbered metal ring (British Trust for Ornithology, BTO, ringing scheme) placed on one leg. Each bird was also fitted with two plastic colour rings on the other leg, the colour code, which could be read using binoculars, identifying the age of the bird, i.e. adult, the wood and the year of capture. To assess the source of replacement birds, i.e. whether reared on site ("home-grown") or immigrant, as many as possible of the young reared in each wood each season were also marked. Birds ringed as nestlings were given BTO rings only, fledged juveniles were also colour ringed, the code identifying them as juveniles of a particular year and the natal wood. The ideal aim was to mark all the adult breeders of certain species and their young each year, but for various reasons, including within season turnover of adults, this was not always possible.

In order to assess the magnitude of movements of individuals between adjacent small woods, several of the study woods were selected as groups. One group comprised five woods within an area of 1.6 x 0.6 kms, and a second group included four woods within an area of 1.0 x 0.4 kms. A further two woods were 1.6 km apart.

The species selected for marking were those which occurred most commonly as breeding species in small woods in our study area and which could be trapped using mist nets. These species were Wren *Troglodytes troglodytes*, Dunnock *Prunella modularis*, Blackbird *Turdus merula*, Robin *Erithacus rubecula*, Blue Tit *Parus caeruleus*, Great Tit *Parus major* and Chaffinch *Fringilla coelebs*.

Results

Between 1990 and 1991, the loss and replacement levels of adult breeders was high for all species, and varied significantly between species (Kruskal-Wallis $H = 17.35$, d.f. = 7, $P = 0.016$) (Table 1 and Figure 1). The proportion of individuals of each species surviving to breed again in 1991 was not correlated with body size ($r = 0.133$, NS).

Table 1. Numbers of breeding individuals present, and marked, in all 16 study woods and percentage breeding in both years.

	WR	RB	CH	DN	BB	GT	BT
Total no. breeding in 1990	68	48	38	67	56	20	30
Total no. ringed in 1990	52	37	20	57	31	18	26
% 1990 birds breeding in 1991	4	5	15	23	26	33	42
Total no. breeding in 1991	28	34	59	38	18	30	
Total no. ringed in 1991	20	23	21	55	31	18	28

WR: Wren; RB: Robin; CH: Chaffinch; DN: Dunnock; BB: Blackbird; GT: Great Tit; BT: Blue Tit.

Surprisingly, the proportion of study woods within which each species went extinct in 1991 was not significantly correlated with the mean proportion of marked individuals of each species which survived to breed in 1991 ($r = -0.580$, NS). However, in some woods, all the marked birds of a given species disappeared between 1990 and 1991, but a local extinction was not apparent because of rapid replacement by unringed individuals which subsequently bred in 1991. Such an event occurred for all the species studied in at least one wood. Thus in the absence of marked birds, these extinctions would have been hidden. These results were calculated from only those study woods in which all the adult breeders were ringed in 1990 (Table 2). When the hidden extinctions were combined with those observed, the correlation between the proportion of woods within which each species went extinct in 1991 and the mean proportion of individuals surviving to breed in 1991 was significant ($r = -0.734$, $0.01 < P < 0.05$).

Table 2. Number of "visible" and hidden local extinctions in 16 small woods

	WR	RB	CH	DN	BB	GT	BT
No. woods occupied in 1990	16	15	13	16	15	10	14
No. visible extinctions	8	3	1	4	4	2	1
No. hidden extinctions	3	5	2	3	1	2	2
No. woods colonised in 1991	0	0	1	0	0	1	1
No. woods occupied in 1991	8	12	13	12	11	9	14

Abbreviations as in Table 1.

Of 103 "homegrown" juveniles of Wren, Dunnock, Blackbird and Robin ringed in 1990, only one (1.0%), a Wren, bred on its natal site in 1991 (and has bred each year since including 1994). A single "homegrown" Robin was present in a different wood in February, but disappeared prior to breeding and was replaced by an unringed bird. A further Robin was present in a third wood in early April in the company of one of the adult breeders from

1990, but both birds had disappeared by May without evidence of breeding and were not replaced. Thus the large majority of replacement breeding birds in 1991 were immigrants.

Within the two groups of woods, 11 birds (six Great Tits, two Blue Tits, two Chaffinches and one Wren) were recorded to have moved between woods (distance moved = 0.2-0.7 km), but these movements did not involve a change in breeding wood either within or between years. A further two birds (a Robin and a Blue Tit) were found breeding in different woods in 1990 and 1991, the distances between the two woods for each bird being 0.4 and 0.5 km respectively. The Robin had previously been ringed in January 1990, at a site 3.5 km from the wood in which it bred in that year. No movements were recorded between the two woods 1.6 km apart for 1990 and 1991, but a Great Tit which bred in one in 1991, bred in the other in 1992.

Discussion

The high rates of turnover of the individual breeders and the concomittant local extinctions showed the bird populations of our small woods to be unstable. In the absence of immigration, the local populations of the species studied would have largely disappeared between one breeding season and the next. The contribution of "homegrown" juveniles to species persistence on their natal sites was negligable. Although birds moved readily between the grouped small woods, most of these movements were not associated with individuals breeding in different woods in consecutive years. Given the high mobility of birds and their potential to disperse over several kilometers, it was not surprising that the majority of replacement breeders appeared to come from sources beyond the immediate locality.

Local extinctions and recolonisations are fundamental to metapopulation dynamics and both processes were recorded in the bird populations of our study woods. However, the high rates of loss and replacement and the distances over which birds were recorded to have moved were more suggestive of a single population with good dispersal ability than off a Levins type metapopulation (Harrison, 1991). Evidence for the latter has been presented for some bird species living in woodland fragments, notably the Nuthatch *Sitta europaea* (Verboom *et al.*, 1991) in which dispersal distances are short. Within our study area, Nuthatches are largely absent, even from suitable habitat in some of the largest woods (*c.* 150 ha.). It is possible that woodland fragmentation in west Cambridgeshire and south Lincolnshire has progressed beyond the stage at which woodland patches are sufficently close to each other to allow a relatively poor disperser to maintain a metapopulation. Other species which do breed in our study area and which are more likely to show metapopulation dynamics are Treecreeper *Certhia familiaris*, Long-tailed Tit *Aegithalos caudatus* and, in particular, Marsh Tit *Parus palustris* which is noted for its sedentariness. None of these species breeds in our small study woods, but reasons for their absence may include habitat and/or area requirements and regional abundance as well as dispersal characteristics.

The high replacement rates of adult breeders were suggestive of high adult mortality. For all species except Blue Tit, the loss of ringed individuals between breeding seasons exceeded published mortality figures (Table 3).

Table 3. Comparison of loss of ringed birds with published mortality estimates.

	WR	RB	CH	DN	BB	GT	BT
% loss of ringed birds	96	95	85	77	74	67	58
Published % annual mortality estimates	63	62	33	23a 71b	34c 44d 69d	50	70

Abbreviations as in Table 1. WR: Hawthorn & Mead (1975); RB: Lack (1965); CH: Newton (1972); DN: a = males, b = females, both for a hard winter, Davies & Lundberg (1984); BB: c = Batten (1973), d = min. and max. values, Coulson (1961); GT and BT: Perrins (1979).

However, the loss of ringed birds included an unknown number which survived, but moved elsewhere. It has been suggested that predation rates in small woods may be higher than in continous woodland (e.g. Andrén, 1992; Paton, 1994). Birds which fail in a breeding attempt are more likely to abandon an area than those which succeeed, therefore high predation rates may promote emigration. Analyses of further data, including the age distribution of replacement breeders (i.e. the relative proportions of first-time and older breeders) may shed more light on this problem. High adult mortality and poor breeding success, if true, and the low rates of "homegrown" juvenile recruitment could suggest that small woods were acting as population sinks (Pulliam & Danielson, 1991). However, young are certainly fledged in our small woods and without knowing the fate of these juveniles once they have left their natal woods, the suggestion cannot be supported. Small woods may simply exchange juveniles over distances of several kilometers or more. Mortality during dispersal can be high, and dispersal through an open landscape is probably more dangerous than through continuous woodland (Lens & Dhondt, 1994), thus the relative contributions of small and large woods to the pool of potential immigrants is of considerable interest.

The hidden extinctions exposed by the marking program showed that true local extinction rates in small populations of some species could be more than double the rate apparent in the absence of a marked population. Attempts have been made to use apparent extinction rates in small populations of relatively abundant species to estimate extinction risks for small populations of endangered species (Pimm *et al.*, 1988). Without knowing the contribution of immigrants to population persistence, risks of extinction could be seriously under-estimated; small, endangered populations are unlikely to receive many, if any, immigrants.

These results are based on data collected during 1990 and 1991. Bird population levels generally in eastern England were reduced in 1991 compared to 1990 following a period of severe weather in February 1991. This regional change was reflected in the decrease in numbers of birds breeding in our small woods in 1991 (Table 1). In subsequent years, a higher proportion of ringed birds have survived between seasons, but the analyses of these data are still in progress.

References

Andrén, H. (1992) Corvid density and nest predation in relation to forest fragmetation: a landscape perspective. *Ecology* **73:** 794-804.

Batten, L.A. (1973) Population dynamics of suburban Blackbirds. *Bird Study* **20:** 251-258.

Coulson, J.C. (1961) The post-fledging mortality of the blackbird in Great Britain. *Bird Study* **8:** 89-97.

Davies, N.B. & Lundberg, A. (1984) Food distribution and a variable mating system in the Dunnock, Prunella modularis. *Journal of Animal Ecology* **53:** 895-912.

Fahrig, L. & Merriam, G. (1985) Habitat patch connectivity and population survival. *Ecology* **66:** 1762-1768.

Harrison, S. (1991) Local extinction in a metapopulation context: an empirical evaluation. M. Gilpin & I. Hanski (Eds). *Metapopulation Dynamics: Empirical and Theoretical Investigations*, reprinted from the *Biological Journal of the Linnean Society* vol. **42**. Academic Press, London 73-88.

Hawthorn, I. & Mead, C.J. (1975) Wren movements and survival. *British Birds* **68:** 349-358.

Lack, D. (1965) *The life of the robin*. Collins, Glasgow.

Lens, L. & A.A. Dhondt (1994) Effects of habitat fragmentation on the timing of Crested Tit *Parus cristatus* natal dispersal. *Ibis* **136:** 147-152.

Levins, R. (1970) Extinction. M. Gerstenhaber (Ed.) *Lectures on Mathematics in the Life Sciences* vol. **2**, *Some Mathematical Questions in Biology*, American Mathematical Society, Providence, Rhode Island 75-107.

Newton, I. (1972) *Finches*. New Naturalist 55. Collins, London.

Paton, P.W.C. (1994) The effect of edge on avian nest success: how strong is the evidence? *Conservation Biology* **8**: 17-26.

Perrins, C. (1979) *British Tits*. New Naturalist 62. Collins, London.

Pimm, S.L.; Jones, H.L. & J. Diamond (1988) On the risk of extinction. *American Naturalist* **132:** 756-785.

Pulliam, H.R. & Danielson, B.J. (1991) Sources, sinks and habitat selection: a landscape perspective on population dynamics. *American Naturalist* **137**, supplement: S50-S66.

Rolstad, J. (1991) Consequences of forest fragmentation for the dynamics of bird populations: conceptual issues and the evidence. M. Gilpin & I. Hanski (Eds). *Metapopulation Dynamics: Empirical and Theoretical Investigations*, reprinted from the *Biological Journal of the Linnean Society* vol. **42**. Academic Press, London 149-163.

Verboom, J.; Schotman, A.; Opdam, P. & Metz, J.A.J. (1991) European Nuthatch metapopulations in a fragmented agricultural landscape. *Oikos* **61**: 149-156.

Wilson, E.O. (1992) *The Diversity of Life*, Allen Lane, London.

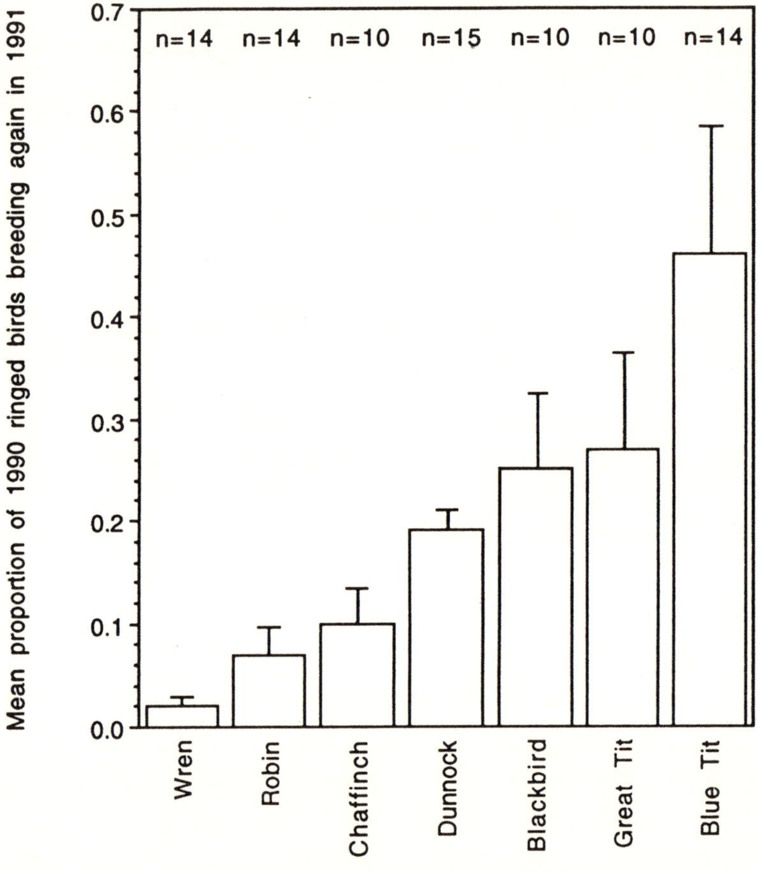

Figure 1. Differences between species in the mean proportions of ringed adults breeding in the same woods in two consecutive breeding seasons (1990-1991). Vertical bars represent + 1 S.E.; n = number of woods for which data were available.

Pheasants and fragmentation: positive effects of landscape interspersion

P.A. Robertson

The Game Conservancy Trust, Fordingbridge, Hants, SP6 1EF, UK.

Abstract

This paper reanalyses previously published data to investigate the effects of habitat composition and interspersion on territorial male pheasant density. Counts of territorial males per km^2 from 147 separate sites were related to the proportions of four habitat types within each area; woodland, arable farmland, grassland and others. Territories were most abundant in areas with 10-40% woodland, 10-40% grassland and 30-80% arable farmland. After removing the effects of habitat composition, the irregularity of the woodland edges within an area explained a significant proportion of the residual variance. Pheasants are clearly a species associated with the intimate interspersion of different habitat types. The consequences for management and the historical effects on British farmland are discussed.

Introduction

The pheasant (*Phasianus colchicus*) has followed man's development of agriculture for over 2000 years. From its origins in China and the northern slopes of the Himalayas this species has been spread by man over much of Europe and North America, becoming a characteristic bird of farmed landscapes. The various introductions of this bird have largely been motivated by its value as a game species, while this same interest has led many individuals and state organisations to manage landscapes for its benefit. Despite this, there has been relatively little work to relate pheasant densities to the composition or broad scale structure of farmed habitats.

There has been a considerable amount of research aimed at quantifying the habitat requirements of the pheasant at different periods during its annual cycle. The birds tend to overwinter as flocks in areas of woodland or other standing residual cover such as reeds or tall, tussocky grasses (Robertson *et al.,* 1993a). In the spring, the males establish exclusive territories along the edges of this cover where it borders onto open ground (Lachlan & Bray, 1976; Robertson *et al.,* 1993b). The males attempt to attract a harem of females, which then feed in the open at dawn and dusk while being guarded by the male (Ridley, 1983). The females leave the male's territory in late April/ May and establish nests. Early nests are typically located in woodland, later nests or renests are often placed in the crops (Hill & Robertson, 1988). The male plays no part in incubation or the care of the chicks. After hatch the chicks are led away from the nest by the female to feed in the growing crops (Hill, 1985). In Britain, pheasants are relatively sedentary, spending the majority of their lives within areas measured in tens of hectares rather than square kilometres.

In this paper I examine the effects of landscape composition and interspersion on territorial male density in the spring, based on a reanalysis of a previously published survey of variations in pheasant density in Britain (Robertson *et al.,* 1993b).

Methods

Breeding pheasant densities were counted on 147 sample areas, each a 1 x 1 km square, across Britain in the springs of 1988 and 1989. Counts were made from a vehicle shortly after dawn or before dusk when the birds were actively displaying on their territories. The habitats within each sample square were then recorded, with particular emphasis being given to the crop contents of individual fields and the shrubby content of the woodlands. The details of the sites, counting methodology and a full listing of the habitat variables recorded in each km^2 are given in Robertson *et al.,* (1993b). For the purposes of this paper it is sufficient to say that the number of male pheasant territories were recorded together with the areas of four habitat types; arable farmland, grassland, woodland and others (including built-up areas, open water, heather moorland and other unsuitable areas for pheasants). In addition, the length of woodland edge was measured and categorised into three categories on the basis of the shrubby cover it contained.

The maximum number of male pheasant territories in a given area appears to be limited by habitat quality. In the presence of more males than available territories, a proportion of males remain non-territorial. Long-term monitoring of spring pheasant densities at one site found that, although total pheasant density varied widely between years it bore no relationship to the number of territories which remained virtually constant. As such, the density of territories can be used as an index of habitat quality (provided that all territories are occupied as can be assumed from the presence of non-territorial males), independent of other factors, such as pheasant rearing, that may influence total bird density, independent of habitat quality (Robertson *et al.,* 1993b).

Data Analysis

An analysis of the effects of habitat fragmentation and interspersion on the density of male pheasant territories must consider two separate issues. Firstly the effects of habitat composition (i.e. the simple proportions by area of each of the four habitat types within each sample square) and then secondly the effects of the distribution of these habitats within the area.

The analysis of data in the form of proportions is not straightforward, the individual proportions cannot be considered independent as, by definition, they sum to one (the unit-sum constraint). To overcome this, a technique known as compositional analysis (Aitchison, 1986) can be employed. This converts the *n* proportions to *n*-1 log-ratios ($\ln(x_i/x_j)$ using the *i*th habitat type as denominator. These log-ratios are then independent of each other and standard multivariate normal statistics can then be applied. This transformation cannot deal with zero proportions for a given habitat type. These were replaced with a small value (0.001), an order of magnitude less than the minimum detectable value (Aebischer *et al.,* 1993) implying that they were present but to a very small extent in each square.

In this case, the initial aim was to identify which habitat types had a significant effect on territory density. The four habitat proportions were converted to six log-ratios using each habitat type in turn as the denominator (*n.b.* only those using the same denominator can be considered independent). An initial examination of the data suggested that the relationship between territory density and the log-ratios was non-linear so the quadratic functions of each log-ratio, together with the cross-products, (see Aitchison 1986 for a full explanation of the procedure) were calculated and entered into a forward stepwise multiple linear regression.

After identifying the relationship between the habitat composition of each square and territory density, the residual variance in the number of pheasant territories was calculated and related to a variety of measures of habitat quality, interspersion and fragmentation. Three measures were used; the number of separate blocks of distinct habitat types (including the full range of arable crops and three categories of woodland separated on the basis of their shrub content); the proportion of all woodland edges with greater than 50% shrubby cover; and an index of the irregularity of the woodland within the area. This last index was calculated as the ratio (ln) of the observed length of woodland edge to that expected if the same area of woodland had been planted as a circle, i.e. the minimum possible length of edge for that area.

Results

Three of the four habitat variables demonstrated significant relationships with territory density: arable farmland, grassland and woodland. The general habitat category 'other' did not appear in the analysis which is unsurprising as it was defined as 'habitats rarely used by pheasants' and was rarely a common feature of the areas counted. The relationship between the three remaining habitat types and territory density is given in Table 1, but is best interpreted by reference to Figure 1. This presents the predicted density of pheasant territories per km^2 for the full range of possible combinations of the three significant habitat types, in each case the area of 'other' was set to zero. Pheasant territories were most abundant in areas with 10-40% woodland, 30-80% arable farmland and 10-40% grassland. The highest densities were found where woodland was present to some extent but the number of territories rapidly declined in areas with over 70% woodland. A mixture of grass and arable farmland was beneficial although a preponderance of arable land over grassland appeared to lead to higher densities.

Figure 1. The predicted density of territorial male pheasants per km^2 in relation to the proportion of woodland, arable farmland and grassland within that km^2. In all cases the area of 'other' habitats has been set to zero. The contours represent increasing numbers of pheasants with the highest density (14 per km^2) occurring near the centre of the figure. The proportion of each habitat type at any point is inversely related to the perpendicular distance from the corner labelled with it's name. Thus the bottom right hand corner is 100% woodland; a point half way along the left hand edge is 50% grassland, 50% arable farmland; while the label '13' on the second innermost contour is 24% woodland, 22% arable and 54% grass.

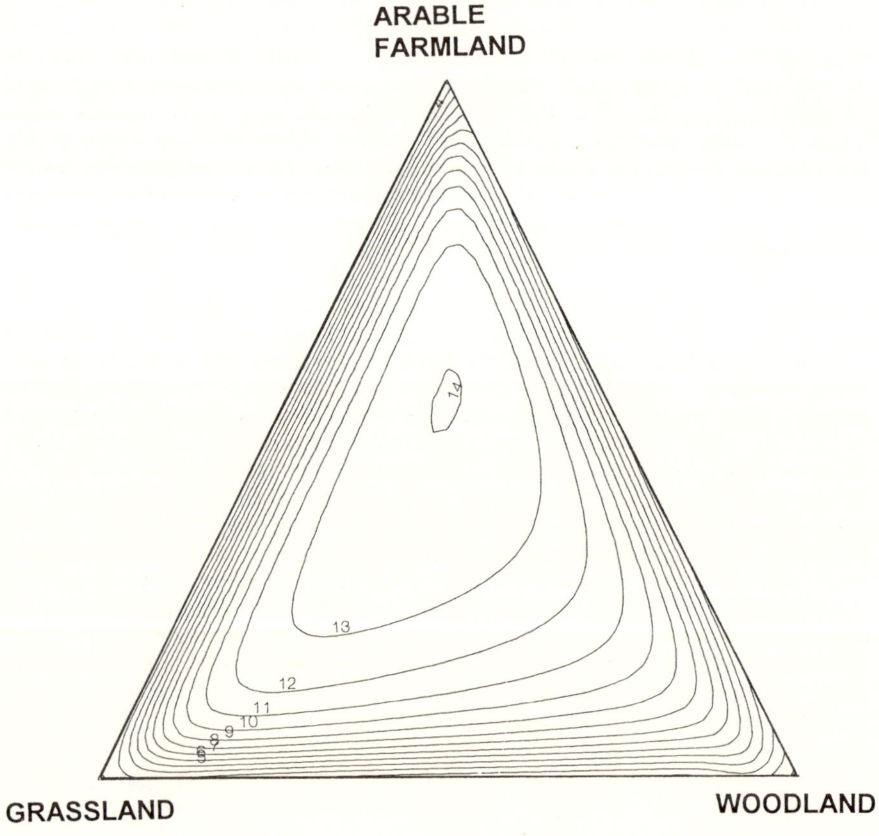

Table 1. Significant variables from a forward stepwise linear regression of the proportion of four habitat types (arable, grassland, woodland and other) per km^2, against the density of territorial male pheasants (ln(x+1)) per km^2. The four habitat proportions were converted to six log-ratios for input into the model, together with the quadratic functions of the log-ratios and their cross products. For the selected variables, A represents ln(arable/woodland) and B represents ln(woodland/grassland). N=149, r^2=0.220, df=3,145, F=13.60, P<0.0001

Variable	Coefficient	s.e.	t	p
CONSTANT	2.66515	0.10840	25.00	<0.001
A	0.07452	0.02235	3.33	<0.002
A^2	-0.03659	0.00648	-5.64	<0.001
B^2	-0.02316	0.01032	-2.24	=0.026

The remaining residual variation in territorial male density was then regressed against the three measures of habitat quality, interspersion and fragmentation. In all three cases there were significant positive relationships between the measures and residual density (no. habitat blocks, r^2=0.032, df=1,146, F=4.927, P=0.028; proportion of shrubby woodland edge, r^2=0.084, df=1,146, F=13.36, P<0.001; edge index r^2=0.101, F=13.65, df=1,121, p<0.001: the latter index could not be calculated for 32 squares that contained no woodland). In a forward stepwise multiple linear regression (p<0.05 to enter or remove), only the edge index was selected.

An overall analysis of the earlier selected variables describing the habitat composition of the area, together with the edge index explained 37.6% of the observed variance in territorial male density between squares. However, in this case one of the quadratic functions (ln(woodland/grassland)^2) was no longer significant (p=0.16). In its absence the remaining variables explained 36.5% of the variation.

Discussion

The pheasant is clearly a species adapted to fragmented landscapes. This makes considerable sense from an ecological viewpoint. The birds are relatively sedentary, spending most of their lives in an area typically less than the 1km^2s used here. They require different habitats throughout the year, woodland as cover during the winter, woodland edges as breeding territories and agricultural crops for nesting and rearing the chicks. As a consequence, the results of the analysis of habitat compositions and the higher densities associated with irregular woodland both reflect the need for the habitats used at different time of year to be placed in close proximity to each other. Although this example is based on the number of territorial males, this is known to be closely related to the number of females in an area although female density is subject to year to year fluctuations in relation to non-habitat related effects such as chick survival in the previous summer and the extent of hand-rearing and shooting (Hill & Robertson, 1988). Given the sedentary nature of the pheasant, spring density per km^2 is also likely to be closely related to the number of hens attempting to nest in the immediate area, although it will bear no relationship to their success. As a consequence, the relationships between territorial male density and habitat are likely to reflect the effects on both sexes throughout much of the year.

Woodland appeared to have a large effect on density. Pheasants are known to concentrate their activities close to the woodland edge (Lachlan & Bray, 1976; Robertson, 1992). The high densities of birds observed in squares with 10-40% woodland and the positive effect of the edge index, a measure of the irregularity of the woodland, both reflect their attraction to areas with a high length of woodland edge per unit area. They are also known to select woodland with a high level of shrubby cover during both the winter and spring (Robertson *et al.,* 1993a,b). The positive effect of the proportion of all woodland edges that contained over 50% cover at a height of 30-200cm reflects this, although this variable was not selected in the final stepwise regression.

There was a lesser effect of the ratio of arable farmland to grassland on territory density, where mixtures of the two, but with arable land predominating, attracted the highest densities. In the late winter and early spring, pheasants graze on the growing shoots of both grass and arable crops. Grass appears to be favoured in early March (Hoodless & Draycott, unpublished) with arable crops predominating in April and May. The early flush of grass in March, when the hens are still in large flocks prior to joining the males on the territories, may help tide them over a period before most arable crops are suitable. When the birds are dispersed and at their most active on their territories in April, both males and females are found at higher densities along woodland edges bordering arable fields as opposed to grassland (Woodburn & Robertson, 1990), presumably as it provides a more nutritious food source at this time. This may explain their higher densities in areas of mixed farming but where arable land predominates.

An important consideration for studying species at a landscape scale is to separate the effects of changing habitat composition from fragmentation as such. The analysis of density in relation to the proportional availability of different habitat types, regardless of their fragmentation or interspersion, is an important consideration, and one faced with considerable statistical problems prior to the development of compositional analysis by Aitchison (1986). Removing the effect of habitat composition allows the effects of fragmentation to be directly and independently assessed.

An appreciation of the landscape-scale requirements of pheasants is of more than academic interest. The management of this species has a considerable effect on the way many farmers and landowners plan their landscapes, in particular the siting, size and shape of their woodlands. This is not just of interest for the management of existing farmlands but, culturally, has influence agricultural landscapes in Britain since the mid-1800's. The role of pheasant management in planning the arrangement of uncropped habitats on farmland is widely recognised by landowners (Cobham Resource Consultants, 1983) and has financial implications for their retention and management in many areas.

References

Aebischer, N. J.; Robertson, P.A. & Kenward, R.E. (1993) Compositional analysis of habitat utilization from animal radiotelemetry data. *Ecology* **74**:1313-1325.

Aitchison, J. (1986) *The Statistical Analysis of Compositional Data.* Chapman and Hall, London.

Cobham Resource Consultants (1983) *Countryside Sports: Their Economic Significance.* The Standing Conference on Countryside Sports, Reading, UK.

Hill, D. A. (1985) The feeding ecology and survival of pheasant chicks on arable farmland. *Journal of Applied Ecology* **22**:645-654.

Hill, D.A. & Robertson, P.A. (1988) *The Pheasant: Ecology, Management and Conservation.* Blackwell Scientific Publications, London.

Lachlan, C. & Bray, R.P. (1976) Habitat selection by cock pheasants in the spring. *Journal of Applied Ecology* **13**: 691-704.

Ridley, M. W. (1983) *The Mating System of the Pheasant Phasianus colchicus.* Unpublished D.Phil. Thesis, University of Oxford. 176pp.

Robertson, P.A. (1992) *Woodland Management for Pheasants.* Forestry Commission Bulletin 106, HMSO

Robertson, P.A.; Woodburn, M.I.A. & D. A. Hill. (1993a) Factors affecting winter pheasant density in British woodlands. *Journal of Applied Ecology* **30**:459-464.

Robertson, P. A.; Woodburn, M.I.A.; Neutel, W. & Bealey, C.E. (1993b) Effects of land use on breeding pheasant density. *Journal of Applied Ecology* **30**:465-477.

Woodburn, M.I.A. & Robertson, P.A. (1990) Woodland management for pheasants: economics and conservation effects. J.T. Lumeij and Y.R. Hoogeveen (Eds.) *The Future of Wild Galliforms in the Netherlands.* Organisatiecommisie Nederlandse Wilde Hoenders, Amersfoort, Netherlands 185-197.

Amphibians and landscape composition

M.J.S. Swan[1] and R.S. Oldham[2]

[1] Amphibian Habitat Advisory Services, 19 St Judith's Lane, Sawtry, Huntingdon, Cambridgeshire, PE17 5XE, UK.
[2] Department of Applied Biology and Biotechnology, De Montfort University, Scraptoft Campus, Leicester, LE7 9SU, UK.

Abstract

Aquatic and terrestrial habitat determinants of three widespread amphibian species were identified using discriminant analysis of data recorded during the NCC National Amphibian Survey. Landscape diversity, in terms of:- (a) intrinsic diversity of individual land-use types and (b) diversity of land-uses, was identified as important in determining species presence. Aspects of metapopulation biology relevant to amphibians in farmland are discussed: juvenile survival, habitat stability and species demography are considered to be particularly important.

Introduction

The widespread British amphibians - common frog, *Rana temporaria,* common toad, *Bufo bufo,* smooth newt, *Triturus vulgaris,* palmate newt, *T. helveticus* and great crested newt, *T. cristatus,* are found in most landscape types within mainland Britain and breed in a wide variety of water-bodies. Nevertheless, despite their relatively non-specialist ecologies even the commonest species, *R. temporaria,* has been recorded in only 47% of systematically surveyed ponds, and the rarest, *T. cristatus,* in only two percent, (Swan and Oldham 1993). Thus, swathes of British countryside are apparently inimical to them. While water-body characteristics associated with each species have been widely investigated, (e.g. Cooke 1975, Cooke & Frazer 1976, Strijbosch 1979), relationships between landscape structure and species distributions are less well understood. Relevant studies have been undertaken by Strijbosch (1980), Beebee (1985), Laan and Verboom (1990) and Mann *et al.* (1991) in which both terrestrial and aquatic variables were included in analyses of habitat determinants. At the population level, Franklin (1993) and Swan (1986) studied terrestrial movements of crested newts and toads in the UK.

In this paper we summarise an analysis of habitat data, collected during the Nature Conservancy Council (now English Nature) National Amphibian Survey (NAS), which identified aquatic and terrestrial factors associated with the presence of common frogs, common toads and crested newts. We have not attempted to provide an exhaustive account of factors affecting the distribution of the species in mainland Britain (the effects of chemical contaminants, for example, are excluded). Rather, in the light of some results from a national volunteer-based survey, we have considered some basic aspects of amphibian metapopulation biology in relation to landscape structure. In particular, we examine the importance to amphibians of diversity of land-use on a local scale.

Methods

NAS methodology is described by Swan and Oldham (1989). Breeding site descriptions provided by volunteer recorders included information on the Amphibia, water-body characteristics and surrounding landscapes within 500 metres of each pond. Only complete datasets for non-flowing, non-garden water-bodies were included in our analyses, providing a sample size of 1,503 records for both anuran species and 1,322 for crested newts whose recorded UK range is smaller (Arnold 1983, Swan and Oldham 1993).

Aquatic variables included site dimensions, extent of shading, vegetation coverage, tendency to desiccate and the presence or absence of fish. Species data were restricted to presence or absence. Each terrestrial element, presented in a tick-list of land-uses and landscape features, was given a score according to its extent and its proximity to the pond. The scoring system is described by Swan and Oldham (1993). This system allowed selection of subsets of data relating to particular predominant land-uses, enabling a two-tier analysis to be undertaken.

Discriminant analysis was used to ascertain whether occupied and unoccupied water-bodies could be distinguished on the basis of the recorded variables. Twenty-two habitat descriptors were entered (listed in Table 1(a), and separate analyses were carried out for each species.

Results

Initially, all records were subjected to stepwise discriminant analysis which identified the most important variables. These were then used in the calculation of the discriminant function (Table 1(a)). Significant discrimination, ($p<0.05$), was obtained in each case. The accuracy of each classification was then assessed using a cross-validation technique (Table 1(b)).

For frogs, seven terrestrial and five aquatic determinants were identified (Table 1(a)). The most important variable was "**arable**", a negative correlate of frog presence. The accuracy of classification was 65% overall, but the model was approximately twice as good at classifying ponds with frogs (81% correct) as ponds without frogs (44% correct). Twelve variables were also selected by the analysis for the toad: aquatic factors ranked highly, indicating that water-body characteristics are important to this species. The accuracy was 74% overall: only 38% for positive sites but 89% for negatives. Analysis of the crested newt data selected three aquatic and five landscape descriptors. Important landscape variables were "**mineral extraction sites**" (positive), and "**flowing water**" (negative). Important positive aquatic variables were "**submerged vegetation**" and "**depth**", but a significant negative correlate was "**fish**". Crested newts in Britain could thus be identified with well-vegetated, fish-free ponds, often in mining or quarrying landscapes. At 89%, the total percentage of sites classified correctly was higher for this than for the other species. However, the accuracy of prediction was far higher for absence of the species (99%) than for its presence (10%).

Table 1. Results of discriminant analysis

(a) List of habitat descriptors and their contribution to the classification of water-bodies according to the presence or absence of each amphibian species. Variables selected by stepwise procedure are labelled "+" or "-" depending on whether they correlated positively or negatively with species presence. Common frog: *Rt*; common toad: *Bb*; crested newt: *Tc*.

AQUATIC VARIABLES	*Rt*	*Bb*	*Tc*	TERRESTRIAL VARIABLES	*Rt*	*Bb*	*Tc*
area				**hedge**		-	
depth	-		+	**ditch**			
shade	-	-		**flowing water**	+	+	-
emergent vegetation				**wetland**		+	
submerged vegetation	+	+	+	**road**			
fish	+	+	-	**scrub**		+	
desiccation	+	-		**moorland**	+		-
				dune		+	
				mineral extraction			+
				improved grassland		-	
				rough grassland	+		+
				gardens	+		+
				built-up areas	-	-	
				arable	-		
				woodland	+	+	

(b) Significance of discriminant analyses and percentage of sites classified correctly.

	PERCENTAGE CORRECT CLASSIFICATIONS			
species	**total**	**sp. present**	**sp. absent**	**P<**
frog	65	81	44	.0001
toad	74	38	89	.0001
crested newt	89	10	99	.0001

To improve the accuracy of positive predictions and to assess pond occupancy within major land-use types, descriptors of sites in predominantly "**arable**", "**improved grassland**", "**rough grassland**", "**woodland**", "**moorland**", "**garden**" and "**built up**" settings were analysed further. The occurrence of each species in water-bodies in landscapes dominated by these land-uses is summarised in Table 2.

Discriminant analysis was performed on data representing four types of land-use, in which structural diversity and availability of ground cover were assumed to vary in the order "**woodland**" > "**rough grassland**" > "**improved grassland**" > "**arable**". In each case the discriminant function obtained was significant (Table 3). The accuracy of positive predictions was improved for toads and crested newts, but not for frogs, suggesting that frog distribution is related more to immediate pond surrounds and less to the wider landscape than that of the other species.

Table 2. Percentage occupancy of ponds in areas dominated by different land-use types.

land-use	**common frog**	**common toad**	**crested newt**	**N**
woodland	61	32	11	230
rough grassland	58	29	14	312
improved grassland	52	13	10	240
arable	39	27	8	51
built-up area	50	20	12	50
gardens	67	24	20	37
moorland	80	26	2	41
all sites	58	28	13	1,503

Table 3. (a) Contribution of habitat descriptors to the classification of water-bodies according to the presence or absence of each amphibian species in landscapes dominated by four major land-use types.

HABITAT	WOODLAND			R.GRASSLAND			I.GRASSLAND			ARABLE		
VARIABLES	*Rt*	*Bb*	*Tc*	*Rt*	*Bb*	*Tc*	*Rt*	*Bb*	*Tc*	*Rt*	*Bb*	*Tc*
area												
depth		+			+	+		+	+			
shade			-								-	
em. veg.												
sub.veg		+	+		+	+	+		+			
fish		+			+			+		+	+	
desicc.		-			-			-				
hedge	+	+			-		-	-	-	-	+	-
ditch	+	-	-			-	+		+		+	+
f.water	+	+			+	-		-	-	+	+	
wetland			+		-	-	-	+	-	-	-	
road	+	-			+					-		-
scrub	+	+			+						+	-
moor								-				
dune												
min.ex.						+						
I.grass			+		-						-	+
R.grass		+								+	-	
garden								+		+	+	+
built-up		-	+	-		+	+	+	+	-	-	
arable		+	+	-	-	+	-		+			
woodland				-	+	+	-	+		+	+	+

(b) Significance of discriminant analyses and percentage of sitesclassified correctly.

p<	0.012	0.001	0.003	0.050	0.001	0.001	0.018	0.001	0.023	0.001	0.001	0.003
	% correct classifications											
total	59	74	73	58	74	79	60	78	66	88	87	94
sp. present	55	68	68	63	71	58	55	71	70	90	93	100
sp.absent	65	77	73	51	79	83	65	79	65	87	84	93

Predictions for all three species were most accurate in sites whose surroundings were predominantly arable, the land-use assumed to have the lowest inherent diversity. Here, terrestrial descriptors assumed the greatest importance relative to aquatic ones as determinants of species presence (Fig. 1). Within more diverse landscapes, specific determinants may be more difficult to identify.

Fig. 1 Relationship between number of significant terrestrial species determinants and number of aquatic species determinants in different predominant land-use types.

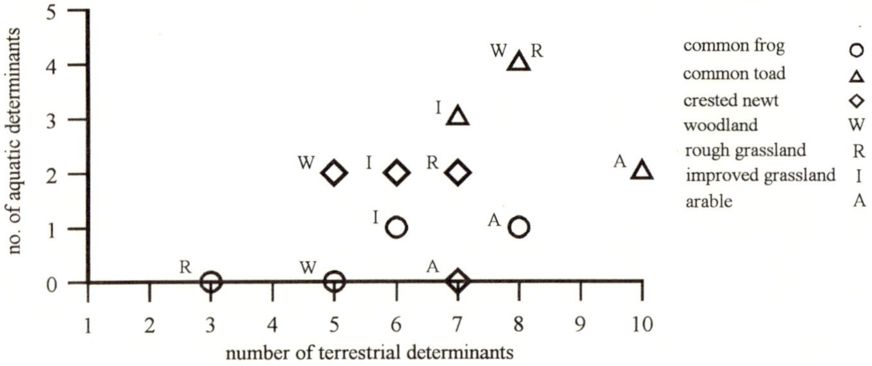

Discussion

All three species were found in each land-use type but, in general, higher proportions of water-bodies were occupied in land-uses considered to have the highest diversity at ground-level. Ponds in landscapes dominated by low-diversity land-use (i.e. arable) had low occupancy rates but were more likely to support amphibians where other land-use types or features were present within 500 metres.

Terrestrial variables can be divided into land-use types (extensive) and landscape features (local). The former tended to be positive species determinants when they occurred within areas dominated by highly managed land, such as arable or improved grassland. Interpretation of relationships between species presence and landscape characteristics may be complicated by co-linearity between variables. For example, in the "**improved grassland**" dataset, "**hedges**" correlated negatively with pond size ($p<0.05$), possibly explaining why the analysis identified a negative relationship between the presence of hedges and the presence of toads. It seems likely that species occurrence may be explained more satisfactorily by particular combinations of variables than by sums or products of variables considered individually, as also reported in Capen (1981). It is possible that such combinations are indicative of agricultural regimes as well as specific landscape structure.

Amphibians persist as metapopulations, groups of discrete populations whose persistence depends on exchanges of individuals between breeding sites (Reh and Seitz 1990, Sjögren 1991). Metapopulation persistence requires sources of migrants, and traversable landscapes through which dispersal can occur (Fahrig and Merrian

1994); i.e. ponds that function as breeding sites, set in landscapes that:- (a) sustain adults, and facilitate their annual breeding migration and (b) allow the dispersal of a proportion of newly metamorphosed animals between water-bodies and sustain their development.

It is usual to consider amphibian lives in two distinct phases, aquatic and terrestrial, but it is appropriate to sub-divide the latter into "juvenile" and "adult". In these three species, post-metamorphic juveniles (metamorphs) comprise the dispersal phase, adults showing strong breeding site fidelity, (Oldham 1985, Reading *et al.* 1991). The concept of "corridors", however, as channels for dispersal may have limited relevance to amphibians in agricultural land. Although a degree of adult fidelity to breeding migration routes has been demonstrated, most agricultural terrain is traversable by adult amphibians in suitable weather conditions (Swan 1986). Migrating adults require refugia en route, enabling them to survive periods of drought or cold, but not specific habitat through which to move (van Gelder *et al.* 1986, Oldham and Swan 1992). Metamorphs disperse from natal ponds rapidly and in all directions (Oldham 1985); this behaviour may result in extensive tracts of barren landscape exacting a high toll and functioning as population sinks. Metapopulation theory states that because individual populations inevitably become extinct without "rescue" by dispersal, so also will metapopulations (Hanski 1991). Elements which enhance the survival of dispersing metamorphs may therefore be critical terrestrial determinants of amphibian metapopulation persistence in agricultural landscapes.

The timing, frequency and extent of agricultural disturbances may also be direct determinants of species complement. Habitat stability varies between agricultural regimes. Within landscapes dominated by permanent pasture or woodland, there may be little annual variation, while mixed, rotational farming areas can undergo considerable land-use changes between years but still maintain relative stability of composition in the longer term. Within such variable mosaics, refugia in non-ploughed areas can be within amphibian walking distance of any point in the landscape. Blanket arable landscapes, on the other hand, may undergo only small year-to-year changes, but every field will be subject to the cycle of ploughing and harvest: i.e. will suffer extreme perturbations at least twice every year from which animals are unlikely to escape in the absence of refugia. The timing of harvest and ploughing operations may be critical. Unfortunately, these tend to coincide with metamorphosis, toads and crested newts being particularly at risk. Thus, the landscape is at its most inhospitable during a critical period.

We have referred to amphibians in a general sense throughout much of this discussion. However, although many aspects of landscape ecology are broadly applicable to all three species, the species exhibit differential resilience to habitat perturbations. In particular, the potential for population recovery and site recolonisation may be important determinants of species survival within rotational landscapes. Frogs and toads are very fecund, producing 1,000-4,000 ova per female per year (Cummins 1986, Swan 1986), therefore populations have the potential to recover rapidly from occasional recruitment failure. However, individuals are relatively short-lived (3-5 years) and most breed only once (Gibbons and McCarthy 1983, Gittins *et al.* 1982). Thus, recruitment failure for four consecutive years could result in extinction. Crested newts, on the other hand, are much less fecund - 100-200 viable eggs laid per female per year (Hedlund 1990) - but longer-lived, and may breed several times. Thus, provided that the level of cover is sufficient to support adults, recolonisation of reinstated water-bodies or terrestrial habitats is possible, even after several consecutive years of recruitment failure. Dispersal ability may therefore be less important to crested newts than to frogs or toads in rotational, fragmented agricultural habitats. Modellers have identified rates of landscape change and the degree of habitat isolation as more important determinants of population survival than demography (Fahrig and Merrian 1994, Hanski 1991, Sjögren 1991), but we suggest that species' demographies can affect the outcome of temporary deleterious habitat change.

References

Arnold, H.R. (1983) *Distribution maps of the amphibians and reptiles of the British Isles.* Biological Records Centre, Abbots Ripton

Beebee, T.J.C. (1985) Discriminant analysis of amphibian habitat determinants in South-East England. *Amphibia-Reptilia* **6:** 35-43

Capen, D.E. (1981) *The use of multivariate statistics in studies of wildlife habitat.* USDA Forest Service General Technical report RM-87, Vermont

Cooke, A.S. (1975) Spawn site selection and colony size of the frog (*Rana temporaria*) and the toad (*Bufo bufo*). *Journal of Zoology (London)* **175:** 29-38

Cooke, A.S. & Frazer, J.F.D. (1976) Characteristics of newt breeding sites. *Journal of Zoology (London)* ***178:*** 223-236

Cummins, C.P. (1986) Temporal and spatial variation in egg size and fecundity in *Rana temporaria. Journal of Animal Ecology* **55:** 302-316

Fahrig, L. & Merrian, G. (1994) Conservation of fragmented populations. *Conservation Biology* **8**: 50-59

Franklin, P.S. (1993) *The migratory ecology and terrestrial habitat preferences of the great crested newt, Triturus cristatus, at Little Wittenham Nature Reserve.* Unpublished MPhil Thesis. De Montfort University, Leicester.

Gibbons, M.M. & McCarthy, T.K. (1983) Age determination of frogs and toads (Amphibia, Anura) from North-western Europe. *Zoologica Scripta* **12**: 145-151

Gittins, S.P.; Steeds, J.E. & Williams, R. (1982) Population age-structure of the common toad (*Bufo bufo*) at a lake in Mid-Wales determined from annual growth rings in the phalanges. *British Journal of Herpetology* **6:** 249-252

Hanski, I (1991) Single-species metapopulation dynamics: concepts, models and observations. *Biological Journal of the Linnean Society* **42:** 17-38

Hedlund, L. (1990) *Reproductive ecology of crested newts, Triturus cristatus (Laur.)* Unpublished PhD Thesis. Swedish University of Agricultural Sciences, Uppsala

Laan, R. & Verboom, B (1990) Effects of pool size and isolation on amphibian communities. *Biological conservation* **54:** 251-262

Mann, W., Dorn, P. & Brandl, R. (1991) Local distribution of amphibians: the importance of habitat fragmentation. *Global Ecology and Biogeography Letters* **1:** 36-41

Oldham, R.S. (1985) Toad dispersal in agricultural habitats. *British Ecological Society Bulletin* **16:** 211-214

Oldham, R.S. & Swan, M.J.S. (1992) Effects of ingested radio transmitters on *Bufo bufo* and *Rana temporaria. The Herpetological Journal* **2**: 82-85

Reading, C.J., Loman, J & Madsen, T (1991) Breeding pond fidelity in the common toad *Bufo bufo. Journal of Zoology, London* **225:** 201-211

Reh, W. & Seitz, A. (1990) The influence of land use on the genetic structure of populations of the common frog *Rana temporaria. Biological Conservation* **54:** 239-249

Sjögren, P. (1991) Extinction and isolation gradients in metapopulations: the case of the pool frog (*Rana lessonae*). *Biological Journal of the Linnaen Society* **42:** 135-147

Strijbosch, H. (1979) Habitat selection of amphibians during their aquatic phase. *Oikos* **33:** 363-372

Strijbosch, H. (1980) Habitat selection of amphibians during their terrestrial phase. *British Journal of Herpetology* **6:** 93-98

Swan, M.J.S. (1986) *The conservation ecology of Rana temporaria and Bufo bufo in Leicestershire.* Unpublished PhD Thesis, Leicester Polytechnic, Leicester

Swan, M.J.S. & Oldham R.S. (1989) *Amphibian Communities: final report to the Nature Conservancy Council,* NCC Research Report, Peterborough

Swan, M.J.S. & Oldham R.S. (1993) *Herptile Sites. Volume 1: National Amphibian Survey Final Report,* English Nature Research Report No. 38, Peterborough

van Gelder, J.J., Aarts, H.M. & Staal, H.W.M. (1986) Routes and speed of migrating toads (*Bufo bufo* L.): a telemetric study. *The Herpetological Journal* **1:** 111-114

Acknowledgements

Our thanks are due to all the contributors to the NAS for providing the information, and to English Nature for funding the project and for allowing us to use the NAS data. We are grateful also to Gilbert Morrey of De Montfort University for help with the statistics and to Clive Cummins of ITE for helpful discussion

PART III:

POSTERS

Effects of isolation on vertebrate and invertebrate populations living in a fragmented landscape

G.W. Arnold, G.T. Smith, M. Abensberg-Traun, D.E. Steven and D. Fortin

CSIRO Division of Wildlife and Ecology, Perth Western Australia

Introduction

During the last 100 years, the development of agriculture in the south-west of Western Australia has resulted in the extensive destruction and fragmentation of the original woodlands, shrublands and heaths. In the Kellerberrin district of the central wheatbelt, the remnants vary in size from less than one hectare to more than 1000 hectares, the majority are less than five hectares (Arnold and Weeldenburg 1991). Within the agricultural landscape, the remnants form a mosaic with considerable spatial heterogeneity in size and distribution, but only make up two to five percent of the landscape.

The effects of fragmentation on the fauna of the region are different to those in the Old World, where fragmentation proceeded slowly over centuries (Angelstan and Arnold 1993). In the wheatbelt, habitat loss has been rapid, especially in the 1920s and the period 1945 to 1970, when over half the present wheatbelt was developed.

For all species, fragmentation caused a massive decline in population, the remaining population generally being made up of numerous small, isolated sub-populations, which because of their size, are vulnerable to stochastic perturbations. Our knowledge of how species respond to the changed environment is poor.

Methods

Species richness and abundance of individual species in remnants were determined for different groups in several different studies over a 9 year period. Both habitat and biogeographic attributes of the remnants were measured and their influence on species richness and abundance examined. Here we consider only the influence of isolation. The indices of isolation used were; 1. Distance from the remnant to other native vegetation (eg. along roads, fence lines or another remnant), 2. The total area of mature vegetation within $5km^2$ of the remnant (an index of the potential pool of animals) and 3. The number of linkages of native vegetation connecting to the remnant.

Findings

In a study of 29 remnants of gimlet woodland (*Eucalyptus salubris*) (table 1) we found that the accessibility of the remnant (numbers of linkages) was significantly correlated with richness and abundance for more groups than were the other indices of isolation. There were significant values for richness in 5 of 9 groups examined and an abundance of 3 of 8 groups. These results indicate the importance, in this study, of narrow (4–5m) linkages in helping retain diversity in ground living groups of animals.

Table 1 Correlations between number of linkages to remnants of gimlet woodland and richness and abundance in different animal groups.

	RICHNESS	ABUNDANCE
Birds	0.359	–
Lizards	0.861 ***	–
Scorpions	0.605 ***	0.549 **
Mygal spiders	–	0.556 **
Isopods	–0.430 *	–0.466 *
Cockroaches	0.062	–0.098
Termites	0.538 **	0.337
Earwigs	0.001	–0.102
Beetles	–0.289	–0.362
Ants	0.454 *	–0.103

* P<005, ** P<0.01, *** P<0.001.

The gimlet woodland remnants supported very few species of small passerines but the numbers found increased with accessibility. In another study, the numbers of species of birds found in 13 completely isolated small heath remnants was significantly correlated with distance (70–600m) to the nearest native vegetation ($r = 0.731$, $P<0.01$). With increasing distance to the remnants one after another species ceased to be found even though the habitat was suitable.

Analysis of the densities of two species of Kangaroos (*Macropus fuliginosus* and *M. robustus*) in 152 remnants showed that densities were influenced by habitat attributes but also by distance to the nearest remnant, by the area of native vegetation within a 25 km^2 radius and the width of corridors where remnants were linked.

Our findings indicate that a whole range of animal groups are affected by isolation as well as fragmentation.

References

Angelstam, P. & Arnold, G.W. (1993) Contrasting roles of remnants in old and newly impacted landscapes: lessons for ecosystem reconstruction. D.A. Saunders, R.J. Hobbs & P.R. Ehrlich (Eds). *Nature Conservation 3: Reconstruction of Fragmented Ecosystems, Local and Regional Perspectives,* Surrey Beatty and Sons, Chipping Norton, 109–125.

Arnold, G.W. & Weeldenburg, J.R. (1991) The distribution of remnant native vegetation in parts of the Kellerberrin, Tammin, Trayning and Wyalkatcham shires of Western Australia. *Technical Memorandum No. 33*, CSIRO Division of Wildlife and Ecology, 28.

Landscape alterations and conservation of a coastal wetland in Western Greece

P. Dimopoulos[1], G. Dimitrellos[1] K. Vassilakis[1] and Th. Georgiadis[1]

[1] Laboratory of Plant Ecology, Department of Biology, University of Patras, GR- 265 00 Patras, Greece

Introduction

The Acheron Delta belongs to the western chain of wetlands of Greece, along the coasts of the Ionian Sea, where the ratio of extant to extinct wetlands is 1.25 (Psilovikos, 1992). The studied area is located in the SW part of Epirus (39°15' N, 20°29' E), and belongs to the lobate morphodynamic type of Delta. The following ecological units of landscape are differentiated within the deltaic plain : sea, sandy habitats, salt marshes, freshwater and brackish water ponds, rivers, wet meadows, tamarisk and reed thickets, riparian forests, cultivated land, hills, hamlets.

Ecological value

Vegetation

The vegetation units occuring in the Acheron Delta are: 1. Aquatic and marshy vegetation characterized by pleustophytes, submerged rhizophytes, emergent aquatic species; 2. Halophilous vegetation of low diversity in coastal salt marshes; 3. Ammophilous, ammo-nitrophilous vegetation of low diversity and ecological value in restricted sandy areas; 4. Vegetation of wet meadows and riparian forests retaining a medium to high ecological potential although they are found in decline; 5. Halophytic scrub vegetation with *Tamarix* in good ecological condition.

Fauna

The total number of vertebrates - except the avifauna - recorded in the estuary ecosystems and the surrounding vegetation, runs to 62 taxa (19 fish species, 5 amphibians, 20 species of herpetofauna, 18 mammals). Three of the fish species and two of the mammals fall into the categories of Vulnerable and Endangered (Red Data Book, 1992). Concerning the avifauna, 85 different species have been recorded till now, 21 (24.7%) of which are listed in the Annex I of the EC Directive 79/409, 46 (54.1%) are protected according to the International Bern Convention (32/A/83), and 9 (10.5%) are included in the Red Data Book of ICBP. The ecological group of Passerines have a much increased percentage in comparison with waterfowl as a consequence of the extensive cultivation within the Delta (Georgiadis *et al.*, 1994).

Environmental changes

The analysis of a series of aerial photographs (1960, 1984) and the verification in the field according to the present vegetation pattern, illustrated an unexpected regression and loss of natural habitats in Acheron Delta. In 1960, natural habitats covered 43.8 % of the total area (1620 ha) while by 1994 they cover 33.7%, mainly due to the extension of agricultural land facilitated by the construction of an irrigation and drainage network during the 1960's. The total loss over 34 years is 163.8 ha, about 4.8 ha per year. The alterations of the natural vegetation into artificial landscape units are analysed and illustrated in Table 1. In the present landscape distributional pattern the fragmentation of agricultural land into more and smaller fields, and the maintentance of natural hedges between them, must be noted as a positive fact. The main factors of degradation (qualitative changes) and/or destruction of natural habitats (quantitative changes) are the following: drainage - irrigation networks, extension of agricultural land and intensive monocultures, unwise use of agricultural chemicals, fires, grazing, and decline of the forest formations and scrublands in the surrounding hills (increase of phrygana from 19 ha to 35 ha and decrease of macchia from 99.5 ha to 83 ha).

Table 1 Landscape alterations (in ha of different land uses) in Acheron Delta (1960-1994).

	A	B	C	D	E	F	G
1960	12.5	153.3	266.0	160.0	832.2	118.5	77.5
1994	10.5	111.0	197.0	120.0	989.0	108.0	84.5

A: Sand dune, B: Salt marsh & Rush belt, C: Reed and Tamarisk belt, D: Riparian belt, E: Cultivation, F: Macchia & Phrygana, G: Other land uses (roads, ditches, hamlets e.t.c.).

Conservation proposals

Acheron Delta is an important coastal wetland that must be directly protected. Measures must be taken to ensure the conservation and the improvement of the quality of the delta ecosystems, in balance with the regulations of the various land uses proposed in the sustainable management plan currently in progress. The basic trends are the following: a. Incorporation of the Acheron Delta into the EC Regulation 2078/92 which concerns the methods of agricultural production in harmony with the conservation and restoration of the natural environment, b. Ecological management of agriculture, c. Extension of the riparian zone to a minimum width of 50-80 m along the river bank, d. Plantations of *Salix* and *Populus* species between cultivation and the riparian forests.

References

Psilovikos, A. (1992) Prospects for wetlands and waterfowl in Greece. Finlayson, C.M., Hollis, G.E. and Davis, T.J. (Eds). *Managing Mediterranean Wetlands and Their Birds,* Proc.Symp., Grado, Italy, 1991, IWRB Spec.Publ.No.20, Slimbridge 53-55.

Georgiadis, Th. (scient.coordin.) et al. (1994) *Ecological impacts from the agricultural activities and management in three coastal wetlands of W Greece (Kalamas and Acheron Delta, Kalodiki fen)*, ENVIREG and Ministry of National Economy (in press).

Hellenic Zoological Society & Hellenic Ornithological Society (1992) *The Red Data Book of Threatened Vertebrates of Greece*, Athens.

The spring decline in a population of *Apodemus sylvaticus* (L.) in a fragmented habitat

S.E. Ferron and W.I. Montgomery

School of Biology and Biochemistry, The Queen's University of Belfast, Medical Biology Centre, 97 Lisburn Road, Belfast BT9 7BL, UK

Introduction

The annual population cycle of wood mice (*Apodemus sylvaticus*) has been well documented (Montgomery 1989a). An autumn/winter peak in numbers is usually followed by a spring/summer low. Most studies however, have concentrated on populations in areas of continuous woodland. Several studies have suggested (Dickman & Doncaster, 1987; Bauchau & Le Boulengé, 1991) that annual density fluctuations may be more marked in woodland fragments. This poster presents preliminary findings from a study which compares the dynamics of *A. sylvaticus* in continuous areas of woodland with those in isolated clumps during the decrease phase of the annual cycle.

Methods

The study area comprised pastoral farmland with scattered woodland in N. Down, N. Ireland. 12 sites in deciduous woodland were censused every three weeks using grids (48m square) of 25 points with 2 Longworth live traps at each point. 5 sites were in small isolated clumps (<3ha in size) and 7 were in larger areas of woodland. Mice were marked individually by toe clipping, sexed, weighed and released at the point of capture. Population size was enumerated as Minimum Number Known Alive (MNA) on each grid (Krebs, 1966) and instantaneous rate of population change was calculated following Montgomery (1989b). Treatment and time effects were investigated in two way anova with repeated measures over time.

Results

There was an overall decline in MNA during the spring. Numbers in the isolated sites decreased sooner and reached a lower density while numbers in continuous woodland did not peak until early April (Fig. 1). These differences were not significant. Neither was there any treatment effect on the instantaneous rate of change. There were also no difference in the average weight of mice in isolated and continuous woodland. In February, the sex ratio was male biased (1:2.1) in isolated clumps and female biased (1.27:1) in continuous woods. The difference in sex ratio between the treatments was significant (χ^2=6.71, p<0.01). As time proceeded the sex ratio approached 1:1 in both treatments. Throughout the study a higher proportion of males were in breeding condition in the continuous woods. Again this difference was not significant. In contrast, in early May 81% of females were in breeding condition in the isolated clumps compared with 17.4% in the continuous woods (p=0.0005, Fisher exact test) Juveniles were captured in 4/5 of the isolated grids but only in 2/7 wood sites during June when juveniles were first recorded. The results suggest little difference in population decline in continuous and isolated woods, although the timing and intensity of the onset of breeding may be greater in the latter.

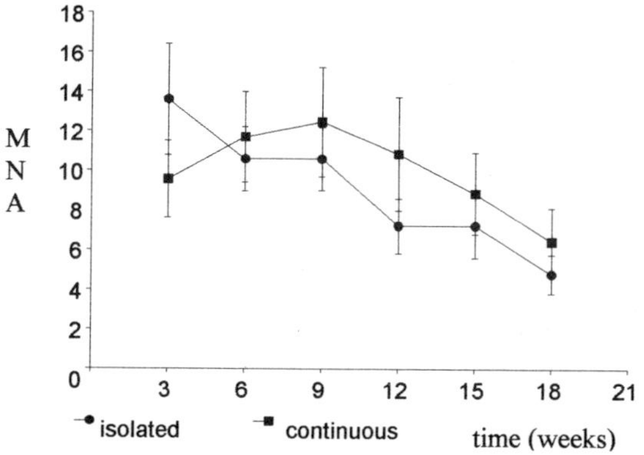

Figure 1
Mean Minimum Number Alive (MNA) for grids set in isolated and continuous woods. Time is shown as weeks from the start of the study, week 0 being 7th February. The first and last sessions were omitted to avoid bias in calculation of MNA.

Conclusions

Sex ratios in the isolated and non-isolated habitat at the end of the winter may be affected by a higher rate of immigration of males into isolated sites during autumn (Ylönen *et al.*, 1991). Female reproduction in *A. sylvaticus* is density dependent (Montgomery, 1989b). The lower density of females in isolated clumps at the end of the spring is probably responsible for an earlier start to breeding. Habitat fragmentation does not seem to have as strong an effect on population dynamics as it does on population structure during the decline phase of the annual population cycle. However, Bauchau & Le Boulengé (1991) found that the greatest differences in dynamics occurred during the autumn and early winter. Habitat quality may also affect population dynamics in isolated woods.

References

Bauchau, V. & Le Boulengé E. (1991). Biologie de populations de rongeurs forestiers dans un escape hétérogène. M.Le Berre & L.Le Guelte. *Le Rongeur et L'Espace* Chabaud, Paris 275-283

Dickman, C.R. & Doncaster, C.P. (1987). The ecology of small mammals in urban habitats: Populations in a patchy environment. *Journal of Animal Ecology* **56:** 629-640.

Krebs, C.J. (1966). Demographic changes in fluctuating population of *Microtus californicus. Ecological Monographs* **36:** 239-273.

Montgomery, W.I. (1989a). Population regulation in the wood mouse *A. sylvaticus.* I. Density dependence in the annual cycle of abundance. *Journal of Animal Ecology* **58:** 465-475.

Montgomery, W.I. (1989b). Population regulation in the wood mouse *A. sylvaticus.* II. Density dependence in spatial distribution and reproduction. *Journal of Animal Ecology* **58:** 477-494.

Ylönen, H., Altner, H-J. & Stubbe, M. (1991). Seasonal dynamics of small mammals in an isolated woodlot and its agricultural surroundings. *Annales Zoologica Fennici* **28:** 7-14.

LANDECONET - An EC funded research programme investigating landscape ecology in changing agricultural landscapes in northern Europe

L.G. Firbank[1], G.L. Fry[2], I. Newton[1], P. Opdam[3], F. Skov[4] and J. Verboom[3]

[1]Institute of Terrestrial Ecology (ITE), Monks Wood, Abbots Ripton, Huntingdon PE17 2LS, UK

[2]Norsk Institutt for Naturforskning (NINA), Postboks 5064 NLH, N-1432, As, Norway

[3]Instituut voor Bos- en Natuuronderzoek (IBN), Postbus 23, NL-6700 AA Wageningen, The Netherlands

[4]National Environmental Research Institute (NERI), Kalo, Grenavej 12, DK - 8410 Ronde, Denmark

---

Introduction

The science of relating landscape variables to biodiversity variables is becoming increasingly prominent for scientific, technical and political reasons. The observational and theoretical basis of landscape ecology is sufficiently developed that it is now clear that the distribution and abundance of some species can be described in terms of variables such as habitat area and degree of isolation. Geographic information systems can be used to integrate the resulting models with landscape data arising from remotely sensed and other sources to help forecast a range of biodiversity measures likely to be associated with a particular landscape. These forecasts can be used, in principle, to devise ways of maximising the benefits to biodiversity within farmed landscapes. Such research is also relevant to the environmental assessment of policies such as the reform of the Common Agricultural Policy and the promotion of the sustainable development of European rural areas.

LANDECONET offers an integrated approach to the landscape ecology of farmland at the European scale. It is a collaborative effort by the European Conservation Institutes Research Network (CONNECT), and is funded for three years from the spring of 1994 under the EC 3rd Framework. The research, co-ordinated by Paul Opdam, comprises a series of international projects with inputs from Belgium, Denmark, Finland, Holland, Norway and the UK.

Empirical studies

Plants (Co-ordinator: Flemming Skov)

This project involves relating the distribution of plants of grassland and woodland to present land disposition and the history of land use. Data will also be collected to relate gene flow in plant populations in particular landscapes to character traits. The objective is to establish predictive rules between landscape pattern and plant diversity.

FRAGMENTATION IN AGRICULTURAL LANDSCAPES

Butterflies (Co-ordinator: Gary Fry)

Here the distribution and abundance of butterflies on farmland will be related to the habitat size, isolation and related variables. The objectives are to determine the response of butterfly species to changes in the landscape pattern, and to assess the degree of exchange of individuals and genes between populations in landscapes of different pattern.

Birds (Co-ordinator: Ian Newton)

Existing studies on the birds on farm woodlands will be developed and integrated. Species richness, turnover, incidence, recolonisation and extinction rates will be related to landscape variables of woodland size and isolation. The degree of regional and annual variation will be established.

Modelling studies (Co-ordinator: Jana Verboom)

Statistical and mechanistic models will be developed using the empirical data to relate biodiversity variables to landscape structure. Models will be developed for metapopulations and to describe individual movement of some species through landscapes.

Application development (Co-ordinator: Les Firbank)

Guidelines for landscape management will be developed, which will include lists of indicator species for particular landscapes. GIS-based application models will be developed which allow the comparison of landscape scenarios in terms of their ecological implications.

Please contact the relevant project leader for further information.

Fragmentation of Grasslands during the last 200 years in S. Sweden

M. Ihse

Department of Physical Geography, Stockholm University, S-106 91 Stockholm, Sweden.

Introduction

Semi-natural grasslands were very important and the ecological base for the agricultural production during several hundred years. They were wide-spread and dominated the landscape in the Nordic as well as in the European countries. During the last fifty to hundred years they successively lost their agricultural importance. Most of their former area has been destroyed. The causes of the loss have been mostly agricultural intensification as ploughing and drainage. Today also agricultural marginalisation as abandoning or afforestation are threats.

Semi-natural grasslands are very important to flora and especially the related invertebrate fauna. They have a great variety of species, but their quality has declined most seriously through fertilisation. This depletion takes place mostly unnoticed.

Methods and materials

A unique possibility to study changes in the landscape long time ago is given in Sweden by historical cadastral maps covering large areas of Sweden in irregular intervals more than two hundred, in some areas nearly four hundred years. In modern times the studies can be done by air-photographs covering all Sweden, from 1940s with photograph intervals of less than ten years, all collected in a central archive.

Historical, cadastral maps have been used to study the changes of grasslands in south Sweden, in the large estate Jordberga, during 200 years with 50 years interval: 1767, 1825, 1875 and 1920 and in air photos with about 10 years interval: 1950, 1967, 1978, 1986 and 1992. The newest air photographs in colour infra red, also give possibilities to map certain aspects of quality of the grasslands. From the old maps also this aspect can be estimated by combining a soil map with an old topographical map where the altitude is mapped in one feet intervals. Three other areas were studied, a river valley and two flat plains, to better understand the pattern and process of fragmentation as compared to the hilly undulating landscape of Jordberga.

Results

In the oldest map from Jordberga, **1767**, the grasslands dominate the landscape. Nearly half the total area is grasslands, mostly connected. Three types of pattern can be distinguished: (a) a biogenetic network with complicated units in patterns connected by corridors, (b) elongated strip along the meandering watercourse and (c) fragmented grassland with patches in small geometric figures.

1767 the **pattern** of grasslands can be explained by the close connection with topography and soils. The distribution follows well the contour lines. The grasslands are mainly found on the slopes and in the small valleys while the ploughed arable field are on top of the hills, where the morain clay soils are dry. The distribution of the grasslands also follows the soils of glacial clay and fen peat. The sloping or low-lying meadows must be mesic or moist. Wet meadows occur on the flooded land around the meandering river.

The grasslands on the flat land was probably dry meadows.

1824, 60 years later, about the same pattern and distribution still exists. The grasslands include small ponds and wetlands, detailed mapped on the map from 1975, especially made for drainage purposes.

1875 the patterns have completely changed. The network of connected grasslands have been dissolved and nearly all the grasslands have been changed to arable fields. Only a few isolated areas remain. The strip along the watercourse is still there but is reduced in size.

1920 a new type of pattern is seen, where the grassland are small point or ring objects found around small ponds and wetlands. This grasslands can be remains from the older meadows surrounding the natural ponds. However, many of the ponds are new artificial marl pits.

This pattern of grasslands in small scattered point/ring objects is still more accentuated in the map from **1950**, the first one interpreted from air photograph, thus giving more detailed information. Probably the same pattern existed in 1920, but the economical map had not the same resolution or aim to map this type of grasslands. In 1950 still one or two small areas existed with continuity from 1767.

In **1992** also this small rings of grasslands around the ponds have been reduced. More than fifty percent of them have disappeared since 1950 as the pond were drained. The grasslands along the watercourse is still more reduced. Grassland vegetation also exists along the linear elements like road verges, stone walls or hedges. The common situation in the plains in southern Sweden is that none or nearly none grasslands with continuity exists today.

The **process** of fragmentation of unfertilised grasslands can be described in four steps with the following patterns:

1. Continuously distributed grasslands over large areas with:
   (a) small scattered agricultural fields on flat land and wide river valleys
   (b) a biogenetic network on slopes and small valleys on hilly land
   (c) elongated strips along watercourses

MI. Many large patches of grasslands surrounded by agricultural fields:
   (a) distributed all over the areas on flat lands, in river valleys and in hilly lands where the connecting corridors have disappeared
   (b) narrow or dissected elongated strips along watercourses.

MII. Few patches with shrinking area.

MIII. Small isolated fragments.

The new grasslands in the agricultural plains are fertilised and ploughed after a few years, mostly less than ten. Fragments of the old grasslands and some of the species can be found in small biotopes like point objects around ponds or linear objects along roads, ditches, stonewalls and hedges. Also among these small biotopes, fragmentation and disappearing occurs. During the last 50 years the large amount of ponds are reduced to a few ones, the linear elements combining an ecological network are dissolved and diminishing with about 50%.

To understand and describe the patterns and distribution of grasslands gives references for evaluating the remaining relics, and knowledge for future nature conservation and restoration. To understand the life strategy of plants in this communities it i necessary to understand the patterns of the grassland as species developed in communities within widespread continuous areas have different demands from those developed in isolated fragmented biotopes.

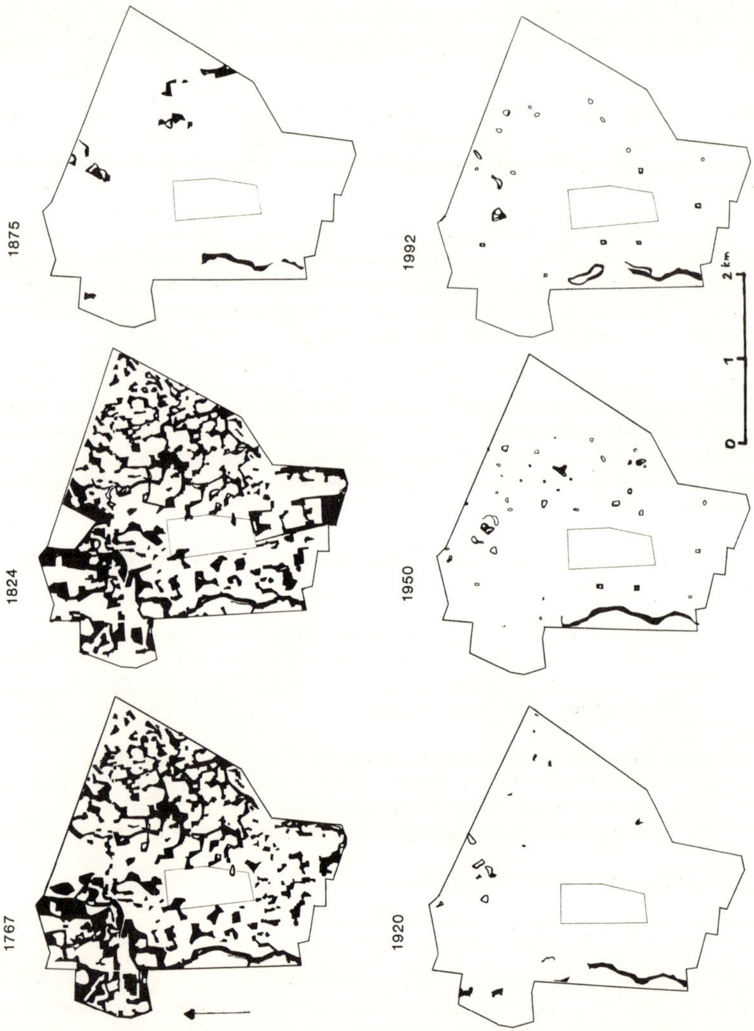

Figure 1. Changes of grasslands (dark) and agricultural fields (white) from 1767 - 1992 in the large estate Jordberga, S.Sweden. The upper row - grasslands mapped from cadastral and economical maps, the lower row - grasslands mapped from air photos.

Fragmentation of woodland by agricultural land

M.C. North

Gillespies, Edmund House, Wood Street, Altrincham, Cheshire, WA14 1ED

Introduction

The species-area ratio as proposed by MacArthur and Wilson (1967) suggests that the number of species present in a given area will be in proportion to that area.

A study of 19 woodlands, ranging from 0.5 to 6 hectares in size, and isolated from each other by agricultural land, has been carried out in the parish of Bold, St. Helens, Merseyside. The number of plant species were recorded for each woodland in order to investigate the validity of the species-area ratio in this case.

Documentary evidence dating from the Domesday Survey of 1086 (Farrer and Brownbill, 1907) makes many references to the woods of Bold, including the planting of woodland in 1730.

Distribution of woodland as shown on the Tithe map of 1840 (L.R.O., DRL 1/11) is not dissimilar to that of today, so these woodlands have been isolated by agricultural land for at least 150 years.

Methods

Species lists for the flora in each of the 19 woods were compiled from existing survey information and personal observations. Flora associated with woodland ponds were excluded from the study. Woodland areas were calculated to the nearest 0.5 hectares from Ordnance Survey maps.

Results

The results of the survey are presented graphically in Figure 1 below.

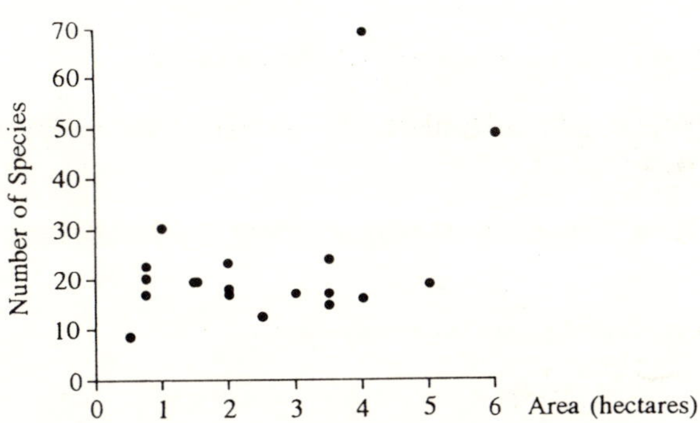

Spearman's Rank Correlation Co-efficient was calculated from the data, and a Student's t-test carried out to ascertain whether any correlation exists between number of species and area of woodland. No correlation was found.

Discussion

Farm woodlands in Bold do not conform to the species-area ratio. Studies by Helliwell (1976) and Middleton and Merriam (1983) on fragmented woodlands show similar findings, although a study by Usher et al. (1992) on farm woodlands suggests a strong correlation between species and area.

The number of species in a woodland can be influenced by age, with older woods having a greater species diversity than younger woods and being characterised by specific plant assemblages (Rackham 1990). However documentary evidence suggests that the woods in Bold are contemporary in age, and non-conformation to the species-area ratio could therefore be used to infer that they were once part of a larger woodland, which has become fragmented by agricultural land.

The Countryside Commission (1983) recognises the landscape and wildlife value of small farm woodlands, and has expressed concern about their future. This study suggests that numbers of farm woodland fragments, together with their degree of isolation and connectivity may be more important in maintaining diversity than fragment size, and it is these issues which must be considered in farm woodland management.

References

Countryside Commission (1983) *Small woods on farms,* CCP 143.

Farrer, W. & Brownbill, J. (1907) *A History of the county of Lancaster 3*, The Victoria History of the Counties of England, London.

Helliwell, D.R. (1976) The effects of size and isolation on the conservation value of wooded sites in Britain, *J. Biogeography 3: 407-416.*

Middleton, J. & Merriam, G. (1983) Distribution of woodland species in farmland woods, *J. Appl. Ecol. **20**: 625-644.*

Rackham, O. (1990) *Trees and Woodland in the British Landscape,* J.M. Dent, London.

Usher, M.B., Brown A.C. & Bedford, S.E. (1992) Plant species richness in farm woodlands, *Forestry **65**: 1-13.*

MacArthur, R.H. & Wilson, E.O. (1967) *The Theory of Island Biogeography,* Princeton University Press.

L.R.O., Lancashire County Record Office, Preston

Influence of scale in the management of fragmented landscapes

M.F. Schmitz, A.J. Rescia, P. Martín de Agar, C.L. de Pablo & F.D. Pineda.

Departamento Interuniversitario de Ecología. Facultad de Biología.
Universidad Complutense. 28040 Madrid, Spain.

Introduction

In order to understand the dynamics of change in a territory, some authors have carried out studies on different scales, with regard to grain and extent (Allen & Hoekstra, 1991). In this context, the analysis of the degree of landscape fragmentation and its spatial pattern is essential. Spatial complexity, for example, allows for the conservation of communities and biological diversity.

The Urdaibai Biosphere Reserve, in the north of Spain, was studied. Here, modifications in the heterogeneity of the landscape have occurred, due to an intense fragmentation process. This work addresses heterogeneity in the context of scale and its relationship with plant diversity.

Methods

Changes in land use were identified on a small and large-scale, analysing the frequency and type of boundaries existing between all the pairs of adjacent uses. This permitted us to calculate an index of spatial complexity -C(s)- (Rescia *et al.*, 1994). Woody plant diversity of the forest masses was also measured (Shannon and Weaver, 1949).The relationship between spatial complexity and the main elements that indicate this, and between these and woody plant diversity on the different scales, were analysed by means of simple and multiple regressions.

Results and Conclusion

Scale-dependent response of the spatial complexity index in the quantification of heterogeneity of the territory

C(s), as an indicating parameter of territory organization, allows one to detect a different evolution in its spatial structure according to the observation scale. The main elements that define spatial complexity of the landscape vary in their degree of contribution to this with change in scale. So, on a small scale, the different types of boundaries (n) generated by the land use units in contact appear as the most significant components of spatial complexity -$R^2(n)=0.81$; $R^2(n+E)=0.92$; $p=0.0001$. On a large-scale the importance of equiprobability (E) is increased in the distribution of the land uses ($R^2(n)=0.53$; $R^2=(n+E)=0.96$; $p=0.0001$, in both cases).

Landscape configuration and plant diversity

On studying the incidence of spatial complexity of the territory in woody plant diversity of forest masses, a significant relationship is only observed on the small-scale (R^2=0.5, p=0.1). The association, also unique to this scale, between the components of the landscape structure and plant diversity is tested: firstly the richness of boundaries (n), R^2=0.6, p=0.007, followed by the evenness (E), R^2=0.5, p=0.1.

From the point of view of territory management, these results indicate that the implementation of conservation and development plans must take into account the type of territory organization and its relationship with the distribution and survival of the species on various scales. In the Urdaibai Biosphere Reserve anything that might provoke modifications in the land use units must consider evidence of these on a small scale (that is, private property) and the effects on the richness of boundaries, as the spatial element which most contributes to the complexity of the territory and is directly related to plant diversity. On a large scale (that is, legal, administrative and economic entities), in order to maintain the spatial configuration of land use units to allow for the conservation of plant diversity, both the richness of the boundaries between uses and their equiprobability distribution must be considered.

References

Allen, T.F.H. & Hoekstra, T.W. (1991) Role of heterogeneity in scaling of ecological systems under analysis. J. Kolasa & S.T.A. Pickett (Eds.) Ecological heterogeneity. Springer-Verlag, New York, 47-68.

Rescia, A.J.; Schmitz, M.F.; Martín de Agar, P.; de Pablo, C.L.; Atauri, J.A. & Pineda, F.D. (1994) Landscape fragmentation, spatial complexity and land management: influence on woody plant diversity in the Basque Country (Spain). J. Veg. Sci, in revision.

Shannon, E.C. & Weaver, W. (1949) Mathematical theory of communication, Univ. Illinois Press, Urbana.